SpringerBriefs in Fire

Series editor
James A. Milke, College Park, USA

More information about this series at http://www.springer.com/series/10476

John Gales • Kathleen Hartin • Luke Bisby

Structural Fire Performance of Contemporary Post-tensioned Concrete Construction

 Springer

John Gales
Department of Civil and Environmental
  Engineering
Carleton University
Ottawa, ON, Canada

Kathleen Hartin
Department of Civil and Environmental
  Engineering
Carleton University
Ottawa, ON, Canada

Luke Bisby
School of Engineering
University of Edinburgh
Edinburgh, UK

ISSN 2193-6595   ISSN 2193-6609   (electronic)
SpringerBriefs in Fire
ISBN 978-1-4939-3279-5   ISBN 978-1-4939-3280-1   (eBook)
DOI 10.1007/978-1-4939-3280-1

Library of Congress Control Number: 2015950062

Springer New York Heidelberg Dordrecht London

Printed on acid-free paper

Springer Science+Business Media LLC New York is part of Springer Science+Business Media (www.springer.com)

Acknowledgments

The authors wish to acknowledge Dr. Martin Gillie and Dr. Tim Stratford for their previous contributions, Mr. Michal Krajcovic for assisting with heating operations, as well as Dr. Florian Block and Dr. Stephen Welch for their intellectual contributions. The Natural Sciences and Engineering Research Council of Canada is recognized for financially supporting this project and Dr. Gales from 2011 to 2013 and for supporting the research training of Ms. Hartin in 2015. The Ove Arup Foundation is also acknowledged for their support of Prof. Bisby during the period in which the research presented in this book was performed. The organizers and participants of COST TU0904 for the Young Researchers Training School for Fire Engineers are recognized. In-kind support is acknowledged from Bridon and Con-Force, UK.

Abstract

Abstract Post-tensioned (PT) concrete is excellent in providing optimized material use for stringent sustainability and aesthetic objectives in modern structures. The continual optimization of PT concrete structures combined with the ongoing development of stronger concretes has resulted in structures which require more engineering attention. This Springer brief intends to provide a more complete understanding of the structural and thermal response of contemporary PT concrete structures to fire. Such an understanding is needed to help practitioners and researchers develop defensible fire-safe designs for these structures and identify current knowledge gaps. Chapters on the following subjects pertaining to PT concrete are included: state-of-the-art summary of contemporary construction, review of previous structural fire test research programmes and real fire case studies, overview of the recent research programme conducted by the author(s), and a summary of current research needs.

Keywords Post-tensioned concrete structures • Standard fire test • Restraint in fire tests • Concrete spalling in fire • Real fire case studies

Contents

Chapter 1
Introduction to Contemporary Post-tensioned Concrete and Fire

*Today's flat-slab post-tensioned buildings...with columns
spaced 12 m on center and span-depth ratios of 40, are more
complex and require more engineering attention than typical
flat-slab buildings of 40 years ago, with columns spaced
at 6 m on center and span-depth ratios of 20.*

—Randall Poston and Charles Dolan (2008)

1.1 Background

In contemporary buildings, less is more. With new construction technologies and
building materials, structures are being built with open floor plans and larger than
ever before spans. Post-tensioned (PT) concrete is excellent at optimizing material
usage, which helps designers meet stringent sustainability objectives while creating
desirable large open spaces. PT concrete uses high strength cold-drawn prestressing
steel tendons that are tensioned inside ducts in the concrete after casting. This com-
presses the concrete prior to loading and results in excellent control of in-service
deflections which is far superior to that of conventional (non-prestressed) reinforced
concrete. This tensioning process results in secondary support reactions in structural systems that balance deflection from loading. The post-tensioning process
reduces the use of building materials enabling large spans. Typical PT concrete
structures include bridges and buildings.

Post-tensioned concrete exists in two forms: unbonded and bonded. In the
unbonded form the prestressing steel tendon is greased and sheathed within the
concrete, preventing any bond to the surrounding concrete. Whereas in the bonded
form, the prestressing steel tendon is grouted within a steel or plastic duct after
tensioning and thus continuously bonded to the concrete.

© The Author(s) 2016
J. Gales et al., *Structural Fire Performance of Contemporary
Post-tensioned Concrete Construction*, SpringerBriefs in Fire,
DOI 10.1007/978-1-4939-3280-1_1

Fig. 1.1 The Shard, an example of a contemporary PT concrete structure

While post-tensioning has been used for some time, as it is now aided by higher strength concretes and computer aided structural optimization, PT concrete contemporary construction has become increasingly complex. Buildings using this technology commonly reach 60 storeys or more. An example being the Shard in London UK (see Fig. 1.1). These advances in contemporary PT concrete construction have changed the way our buildings are designed, yet we have not substantially changed the way we look at them in case of fire. These structures now require more engineering attention and understanding to guarantee safety in the case of fire. Prescriptive building code requirements, particularly in the United States, have failed to keep up with innovation, with no significant changes for many decades. The quote by Randall Poston (Chair of American Concrete Institute committee 318, 2011), and Charles Dolan at the start of this chapter presumably refers to the ambient temperature structural design of PT concrete structures; however, it can just as easily be considered to apply to issues around structural design for fire.

1.2   The Fire Problem

In practice, a key benefit for designers of PT concrete structures, opposed to those of steel structures, is their widely claimed 'inherent' fire endurance. Starting as early as the 1950s, confidence in PT concrete's ability to resist fire was 'demonstrated' through a series of experiments conducted under standard fire conditions in large scale testing furnaces; however, the applicability of these standardized fire tests to real structures, with real fires, is debatable on various grounds (see Bisby et al. 2013). Standard fire tests fail to capture many of the complexities of real structural response of PT concrete structures in real fires. For example, when unbonded and stressed prestressing steel in a floor runs continuously through several bays of a building, local damage to the tendon will affect the structural response and capacity of adjacent bays. In cases where there is not additional conventional bonded reinforcement, as permitted by some codes, the PT concrete slab could lose tensile reinforcement across multiple bays, including areas not directly exposed to the fire. A standard fire test of a simple span is incapable of capturing this global structural response. Additionally, since most fire tests of PT concrete slabs are more than 50 years old, the materials and construction techniques used in available tests are not representative of those used in contemporary PT concrete construction (span-to depth, high strength concretes, etc.). To date, no structural fire tests have realistically demonstrated the degree to which restraining mechanisms in continuous PT concrete floor slabs can be created and sustained during a fire to prevent collapse.

Prestressing steel tendons are more sensitive to high temperature damage than mild steel reinforcement, and therefore require more concrete coverage for protection from fire. Spalling is the loss of concrete cover during and fire. It is due to multiple factors including moisture content, high stress states, and high temperatures. If spalling occurs the prestressing steel tendon can become directly exposed to severe heating. This is an issue for both buildings and bridges, and affects the performance of PT concrete structures in fire has been the subject of debate within the structural engineering community for some time (e.g. Bailey and Ellobody 2009; Kelly and Purkiss 2008). Clearly, a more complete understanding of the structural and thermal response of PT concrete structures in fire is needed. This book investigates some of the potential problems that contemporary PT concrete structures may face in real fires. It is intended to guide current practice and future research towards the development of defensible performance based and fire safe designs.

1.3 Objectives

There are a number of load carrying mechanisms and failure modes for PT concrete structures in fire that have not yet been identified or have not been adequately considered in literature. In particular, because PT concrete floor structures can be continuous across multiple bays, fire tests on isolated simply-supported members with short tendon lengths cannot be considered as representative. Tendon continuity, deflections, and restraint can all be expected to play significant roles in real structures. The research presented in this book aims to:

- explore the current knowledge available on PT concrete structures from traditional experimental fire tests and the global behavior real PT buildings exposed to real fire;
- identify and highlight the structural actions, such as thermal bowing, restraint, concrete stiffness loss, continuity, spalling, and slab splitting, that are likely to occur in continuous, multi-bay PT concrete structures in fire; and
- study and detail a sensor rich research program that considered idealized PT concrete structures conducted for this book, with the intent of providing credible and detailed data for future validation of computational models that can predict the response of PT concrete structures during fires, thus enabling defensible performance based structural fire design.

1.4 Outline

The following is a brief outline of each chapter for this book.

Chapter 2: *Contemporary Post-tensioned Concrete Construction* presents the current practice and fire design of PT concrete structures. Over the last 60 years, the complexity of PT concrete construction has been increasing beyond the scope of the current design guidance with respect to fire safety.

Chapter 3: *Structural Fire Test Research Programmes for Post-tensioned Concrete* evaluates the structural fire performance of PT concrete structures using information available in peer-reviewed literature as well as case studies of real fires in real buildings. While this chapter pays considerable attention to unbonded PT concrete construction due to its tendency for tendon rupture, a brief discussion of bonded is included. The intent of the chapter is to highlight the existing knowledge on the fire performance of PT concrete structures.

Chapter 4: *Localized Heating of Post-Tensioned Concrete Slabs Research Program* features a recent test program conducted for this book. Three novel large-scale tests on locally heated, continuous PT concrete slab strips were conducted to gain a better understanding the structural actions in contemporary multi-bay, continuous, unbonded PT buildings under fire. When these buildings are exposed to fire, other structural actions (thermal bowing, restraint, concrete stiffness loss, continuity, spalling, slab splitting etc.) may play significant and often inter-related roles in influencing the response. This response is studied and explained.

Chapter 5: *Recommendations for Advancing the Fire Safe Design of Post-Tensioned Concrete* provides a discussion of the previous chapters and identifies knowledge gaps and current research needs for creating rational performance objectives for PT concrete structures in fire.

References

ACI. 318. (2011). *Building code requirements for structural concrete* (Rep. No. ACI 318-11). Farmington Hills, MI: American Concrete Institute.

Bailey, C. G., & Ellobody, E. (2009). Comparison of unbonded and bonded post-tensioned concrete slabs under fire conditions. *The Structural Engineer, 87*(19), 23–31.

Bisby, L., Gales, J., & Maluk, C. (2013). A contemporary review of large-scale non standard structural fire testing (circa 1980-present). *Fire Science Reviews, 2*(1), 1–27.

Kelly, F., & Purkiss, J. (2008). Reinforced concrete structures in fire: A review of current rules. *The Structural Engineer, 86*(19), 33–39.

Poston, R., & Dolan, C. (2008). Re-organizing ACI 318. *Concrete International, 30*(7), 43–47.

Chapter 2
Contemporary Post-tensioned Concrete Construction

*In our fire test specifications we require certain minimum sizes
of specimens in order to have a fairly standard test. These are
comparatively large and represent an average if not maximum
of what will be used in the [real] structure.*

—Simon Ingberg, chief of Fire Resistance Section National
Bureau of Standards and former chair of ASTM E05 as
corresponded to Carl Menzel in relation to his standard fire tests
on concrete members (1943).

2.1 Defining a Post-tensioned Structure

Post-tensioned (PT) concrete is an increasingly popular technology since it allows
for rapid construction using less material than conventional concrete. While its use
has been widespread in the United States since the late 1950s, it has recently seen
wider popularity in Europe, China, and the Middle East. PT concrete uses high
strength cold-drawn prestressing steel tendons. The tendon configurations for PT
concrete are bonded and unbonded. In both configurations, prestressing steel is used
to pre-compresses the concrete slab before loading and results in longer spans without deformation. In comparison to conventional reinforced concrete slabs, PT concrete provides excellent control of in-service deflections (Khan and Williams 1995).
Figure 2.1 illustrates simplified tensioning and compression mechanics used in PT
concrete.

In order to explain how optimization of PT concrete construction is achieved, an
understanding of load balancing is necessary (Lin 1963; Aalami 2007). In load balancing the prestressing tendon, which typically follows a parabolic profile within

© The Author(s) 2016
J. Gales et al., *Structural Fire Performance of Contemporary
Post-tensioned Concrete Construction*, SpringerBriefs in Fire,
DOI 10.1007/978-1-4939-3280-1_2

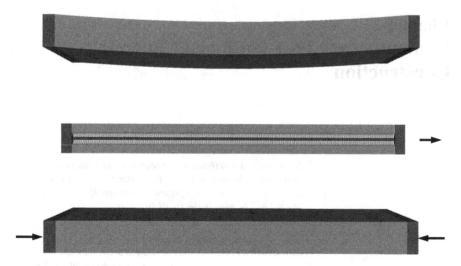

Fig. 2.1 Post-tensioning concrete where the tendon is shown in blue and the forces are shown in *black*

the concrete, acts as a 'distributed force' applied on the entire structural element. This distributed force is a function of the tendon's degree of tension and its drape (the prestressing steel's eccentricity with respect to the neutral axis of the structural element). The distributed force acts to 'balance' the net loading that contributes to service deflection. Figure 2.2 provides a visual representation of load-balancing by tensioning reinforcement.

Figure 2.2 is a highly idealized schematic of loading behavior in PT slabs; in reality the secondary loading exerted by the stressed prestressing steel is more complicated. This complication is partially due to the fixity of support conditions, the compressive reactions at slab ends, and the amount of tendon draped above and below the structure's neutral axis. The decreased deflection in PT structures helps minimize the concrete used by enabling shallower, and optimized structural elements. This enables a designer to specify additional space and open plan compartments for hi-rise buildings.

2.2 Design and Construction Is Changing

Today a typical multi bay, hi rise PT construction consists of a shallow floor (150–200 mm) and long spans of 7–14 m (see Fig. 2.3), creating span to depth ratios of 40 or more. The complexity of this type of building requires special consideration and most guidance available is dated and not be applicable for modern fire engineering design.

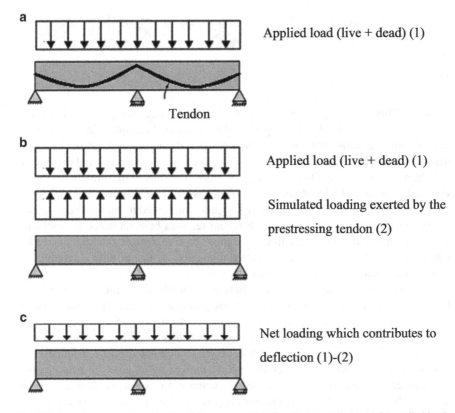

Fig. 2.2 Load balancing after Lin (1963) and Aalami (2007) with (**a**) applied load (**b**) applied load amd simulated loading exerted by prestressing tendon and (**c**) net loading contributing to deflection

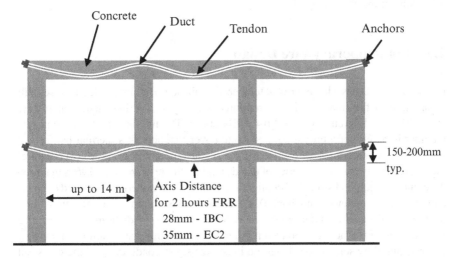

Fig. 2.3 Typical configuration of a multi bay post-tensioned slab

Fig. 2.4 Unbonded tendon continuity

In the 1960s when post-tensioning was in its infancy, designers traditionally relied on simplified load balancing procedures. Today, computer-aided structural design software helps optimize the concrete material usage by calculating the optimal placement and position of tendons. This advancement has permitted designs of flat slabs with span to depth ratios of 40 or more. The tensioning strands have also changed over time. Older structures were cast with isolated monostrand tendons, today's post-tensioned structures are highly pre-compressed using banded tendons, usually in groups of four. These multi strand tendons help reinforce flat slabs to carry primary loads to the structure's columns. To complement the high degree of tensioning and to meet faster construction demands, the use of high strength concrete in PT slabs is becoming more common today.

Beyond changing the number of strands in the tendons, the physical properties have also been altered. Most innovations in prestressing steel have largely been driven by corrosion resistance. For example ACI 318-14 now pushes for full encapsulation of unbonded tendons to help alleviate this concern. Corrosion causing any localized damage can effect adjacent bays in unbonded post-tensioned construction. These concerns have led to changes in the way tendons are being made. Tendons can be alloyed and controlled to increase corrosion resistance, such as through the addition of chromium and copper. Corrosion resistance is especially important in bridges, as stressed tendons in bridge girders are more susceptible. Tendons are now also made to achieve higher strengths. Tendons from the 1970s were typically of 1700 MPa strength, today most tendons are produced to meet a minimum strength of 2000 MPa.

2.3   Contemporary Fire Design

Prestressing steel has a lower critical temperature, the temperature at which it loses half its strength in fire, than mild steel reinforcement. Typically these temperatures are about 100 °C different (ASTM 2014; CEN 2004). The fire design of a PT concrete member is based upon providing a concrete cover that acts as a thermal barrier and protects the prestressing steel from reaching critical temperatures. These cover requirements are largely based on a series of dated standard fire tests from the 1960s that used different materials and structural configurations than their modern constructed counterparts (these are reviewed in Chap. 3). Based on the assumption that concrete is fully restrained, the cover for interior spans is reduced in some jurisdictions. Good practice for fire design is to include a bonded reinforcing mesh to mitigate spalling, although it is not required by some codes based on in-service stress conditions and there is scant

Table 2.1 Concrete axis distance covers as specified in design guidance for floor slabs

| Fire resistance rating (min) | Required fire resistance axis distance (mm) | | | | |
| | Prescribed by EN 1992-1-2 (2004) | | | Prescribed by the IBC (2012) | |
	Simply supported slabs	Continuous slabs	Flat plate slabs[a]	Simply supported slabs[b]	Continuous slabs[b]
30	25	25	25	–[c]	–[c]
60	35	25	30	–[c]	–[c]
90	45	30	40	–[c]	28
120	55	35	50	47	28
180	70	45	60	59	34
240	80	55	65	–[c]	41

[a]15 mm additional axis distance was added to tabulated data from CEN (2004)
[b]Clear cover adjusted to axis distance, by adding 3 mm sheathing and ½ bar diameter of 12.7 mm
[c]The IBC (2012) does not tabulate values for these fire resistances

evidence as to its effectiveness in this regard. Table 2.1 details the prescriptive axis-cover requirements for PT concrete slab systems.

Current cover requirements found in design guidance come from dated experimental test regimes that do not resemble contemporary construction, and limited efforts have been made to develop new guidance that meets the needs of contemporary construction (see Chap. 3).

A comprehensive set of papers (Gales et al. 2011a, b, c) detail the origins of these guidelines. Outlined in these papers were several concerns relating to contemporary PT concrete construction. In particular, floor construction with respect to fire is examined to show why renewed research attention is essential for fire safety in contemporary construction. These concerns are subsequently outlined below.

2.3.1 Tendon Continuity Across Multiple Bays

In unbonded PT (UPT) buildings the prestressing steel tendons are free to move longitudinally through ducts within beams and/or slabs, unlike in the bonded case. Since the same tendon is used across the entire slab, any localized heating and resulting damage has consequences across all bays of the structure.

In PT construction the length of the unbonded tendon can be up to 70 m (Taranath 2010). This is problematic as research has shown that the longer the total length of an unbonded tendon between anchors, the greater the likelihood of the prestressing tendon rupturing due to localized heating (Gales et al. 2011b, c). This is because of a complex combination of accelerating plastic strains and loss of ultimate tensile strength of the tendon. The consequences of premature tendon rupture during fire have received limited research attention to date, despite the fact it has led to the progressive failure of several floors (Post and Korman 2000), and the forced demolition of several UPT structures after real fires (Barth and Aalami 1992; Brannigan and Corbett 2009).

2.3.2 Higher Strength Concretes

Research has shown that high strength contemporary concrete mixes are more likely to experience fire-induced explosive spalling than their low strength counterparts (Kodur and Phan 2007). The satisfactory performance of PT elements in fire resistance tests conducted decades ago and used to define code should not be taken as evidence of adequate fire resistance for modern PT structures, where spalling is likely to occur.

2.3.3 Pre-compression

PT slabs and beams are axially pre-compressed (see Fig. 2.1), which means that a greater portion of the soffit will be subjected to higher compressive stresses in service than in a non-prestressed member. Since compressive stress is a widely acknowledged risk factor for spalling in fire (Hertz 2003), it is reasonable to assume that that PT beams and slabs are more likely to spall.

2.3.4 Large Span-to-Depth Ratios

A primary advantage of PT structures is that they enable span-to-depth ratios greater than ratio of 40 allowed by non-prestressed flooring systems (CPCI 2007). There is a risk that a PT slab could experience proportionally greater deflections during fire, due to thermal bowing caused by uneven heating gradients in the slab and smaller lateral restraint forces. Since compressive membrane action only occurs for a certain level of deflection, there may be less chance of developing this beneficial behavior which provides additional support against collapse. Inability to engage compressive membrane action is a credible concern for PT slabs in fire. This is particularly true in the case of UPT tendons as they can rupture during fire and some design guidelines allow zero mild steel reinforcement in sagging moment regions (CEN 2004; CSA 2004; ACI 2011), see Fig. 2.5.

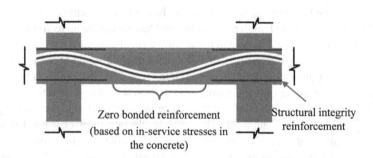

Fig. 2.5 Lack of bonded reinforcement in some service stress conditions

Worth noting is that the span-to-depth ratios used in available furnace tests on UPT members have generally been unrealistically small due to available furnace sizes.

2.3.5 Fire Resistance Based on Concrete Cover

The code prescribed covers are based largely on results of standard fire tests conducted on isolated PT slabs and beams in the 1950s and 1960s (Troxell 1959; UL 1966; Gustaferro et al. 1971), although some codes have been revised based on more recent furnace testing (see Purkiss and Kelly (2008) for discussion pertaining to the development of EN 1992-1-2 (CEN 2004) requirements). In some cases these tests used concrete that had been pre-conditioned before fire testing, did not adequately define the degree of restraint, or used a light steel mesh within the concrete cover to prevent or arrest spalling. Cover spalling was thus explicitly excluded from many of the tests, yet these important details are routinely omitted when using test data to justify current prescriptive cover requirements. Clearly, use of light steel mesh in the cover and pre-drying of the concrete is non-representative of typical PT construction. Additionally, the restraint used in standard fire test furnaces does not always represent a structure's true stiffness, which varies between building configurations and will affect how the structure behaves in fire. In instances where 'total' restraint is achieved, premature failure of a PT structure could be observed due to explosive spalling (see Bletzacker 1967). For these reasons, the wisdom of relying on prescriptive concrete cover thickness as the sole means of achieving fire resistance in UPT buildings is clearly questionable.

2.3.6 Inadequacies of Standard Fire Tests

It is widely recognized that fire exposures used in standard fire tests (e.g. ASTM E119 2014; ISO 834 1999) are unrealistic, particularly for the large compartments or open floor plans found in many modern buildings (Stern-Gottfried et al. 2010). It is generally argued that standard fires are conservative representations of worst-case fires for most types of construction. This rationale cannot be applied to UPT concrete beams or slabs because unbonded prestressing steel tendons are continuous over multiple bays, and localized or travelling fires is more likely to result in premature tendon rupture (Chap. 3 will provide greater details on this issue).

2.4 Summary

The challenges we face in safely designing a PT concrete structure are growing as we use new materials and design technologies to make contemporary structures more complex. The above discussion addresses numerous concerns and inadequacies of current knowledge on the fire performance of PT concrete buildings, particularly with unbonded tendons. To further highlight the apparent disconnect between existing knowledge, available guidance, and industry practice, Chap. 3 summarizes and critically appraises the 46 fire tests on UPT concrete structural members that are currently available in the literature. A brief discussion of bonded PT concrete structural members is also provided.

References

Aalami, B. O. (2007). Critical milestones in development of post-tensioned buildings. *Concrete International, 29*(10), 52–56.

ACI. 318. (2011). *Building code requirements for reinforced concrete* (Rep. No. ACI 318-11). Detroit: American Concrete Institute

ASTM. (2014). *Test method E119-14: standard methods of fire test of building construction and materials* (Rep. No. E119-14). West Conshohocken, PA: American Society for Testing and Materials.

Barth, F., & Aalami, B. (1992). *Controlled demolition of an unbonded post-tensioned concrete slab* (PTI special report). Post-Tensioning Institute, Pheonix AZ, 34 pp.

Bletzacker, R. W. (1967). Fire resistance of protected steel beam floor and roof assemblies as affected by structural restraint. *Symposium on Fire test methods. American Society of Testing Materials*, pp. 63–90.

Brannigan, F., & Corbett, G. (2009). *Building construction for the fire service* (4th ed.). Sudbury, MA: Jones and Bartlett.

Canadian Standards Association. (2004). *CAN/CSA A23.3-04: Design of concrete structures.* Ottawa, ON: CSA.

CEN. (2004). *Eurocode 2: Design of concrete structures, parts 1–2: General rules-structural fire design, EN 1992-1-2.* Brussels: European Committee for standardization.

CPCI. (2007). *Design manual: Precast and prestressed concrete.* Ottawa, ON: Canadian Prestressed Concrete Institute.

Gales, J., Bisby, L., & Gillie, M. (2011a). Unbonded post tensioned concrete in fire: A review of data from furnace tests and real fires. *Fire Safety Journal, 46*(4), 151–163.

Gales, J., Bisby, L., & Gillie, M. (2011b). Unbonded post tensioned concrete slabs in fire – Part I – Experimental response of unbonded tendons under transient localized heating. *Journal of Structural Fire Engineering, 2*(3), 139–154.

Gales, J., Bisby, L., & Gillie, M. (2011c). Unbonded post tensioned concrete slabs in fire – Part II – Modelling tendon response and the consequences of localized heating. *Journal of Structural Fire Engineering, 2*(3), 155–172.

Gustaferro, A. H., Abrams, M., & Salse, E. (1971). Fire resistance of prestressed concrete beams study C: Structural behaviour during fire tests. *PCA Research and Development Bulletin, 29,* 1–29.

Hertz, K. D. (2003). Limits of spalling of fire-exposed concrete. *Fire Safety Journal, 38,* 103–116.

IBC. (2012). *International building code.* Country Club Hills, IL, USA: International Code Council.

ISO 834. (1999). *Fire resistance test – Elements of building construction.* Geneva: International Organization for Standardization.

Kelly, F., & Purkiss, J. (2008). Reinforced concrete structures in fire: A review of current rules. *The Structural Engineer, 86*(19), 33–39.

Khan, S., & Williams, M. (1995). *Post-tensioned concrete floors*. London: Butterworth-Heinemann.

Kodur, V., & Phan, L. (2007). Critical factors governing the fire performance of high strength concrete systems. *Fire Safety Journal, 42*, 482–488.

Lin, T. Y. (1963). Load-balancing method for design and analysis of prestressed concrete structures. *ACI Journal Proceedings, 60*(6), 719–742.

Menzel, C. (1943). Tests of the fire resistance and thermal properties of solid concrete slabs and their significance. In Proceedings of the ASTM (pp. 1099–1153).

Post, N., & Korman, R. (2000). Implosion spares foundations. *Engineering News Record, 12*, 12–13.

Stern-Gottfried, J., Rein, G., Bisby, L., & Torero, J. (2010). Experimental review of homogeneous temperature assumption in post-flashover compartment fires. *Fire Safety Journal, 45*(4), 249–261.

Taranath, B. S. (2010). *Reinforced concrete design of tall buildings*. Taylor and Francis: CRC.

Troxell, G. E. (1959). Fire resistance of a prestressed concrete floor panel. *Journal of the American Concrete Institute, 56*(8), 97–105.

Underwriters Laboratories. (1966). *Report on prestressed pre-tensioned concrete inverted tee beams and report on prestressed concrete inverted tee beams post-tensioned (R4123-12A)*. Chicago, USA: Underwriters' Laboratories Inc.

Chapter 3
Structural Fire Test Research Programmes for Post-tensioned Concrete

The object of all tests of building materials should be to determine facts and develop results that may be of practical value in future designing. In order that such facts and results may have real value, three conditions are necessary: first, that the materials tested shall be identical with what is commercially available in the open market; second, that the conditions, methods, and details of constructions conform exactly to those obtainable in practice; third, that the tests be conducted in a scientific manner.

—Abraham Himmelwright, On the development of a standardized fire test (1898)

3.1 Background

In 1898, in the offices of the Roebling Company at 121 Liberty street in New York City (present day location of the 'Ground Zero' World Trade Center memorial), Abraham Himmelwright wrote the original necessary requirements for ideal structural fire testing to be of value. Structural fire tests performed at Himmelwright's time were based upon an experimental methodology developed by mechanical engineer of the New York Building Department, Gus Henning in 1896 (Himmelwright 1898; Woolson 1902). Inspired to respond to the various ad-hoc structure fire tests of materials, the test series were conducted to develop *"actual and relative efficiencies of different floor constructions"* under fire (Himmelwright 1898). Henning would later denounce the early fire testing as fraud due to its non-representation of reality (Fig. 3.1).

Following this and similar criticism towards the New York building structure fire test series, various construction material agencies lobbied for change (Woolson and Miller 1912). This effort was organized by Ira Woolson at the American Society of Testing of Materials (ASTM 2014). A new fire test standard evolved and was proposed in 1916. Since 1916 the fire testing and engineering community has largely followed that original testing procedure for construction materials and assemblies

© The Author(s) 2016
J. Gales et al., *Structural Fire Performance of Contemporary Post-tensioned Concrete Construction*, SpringerBriefs in Fire,
DOI 10.1007/978-1-4939-3280-1_3

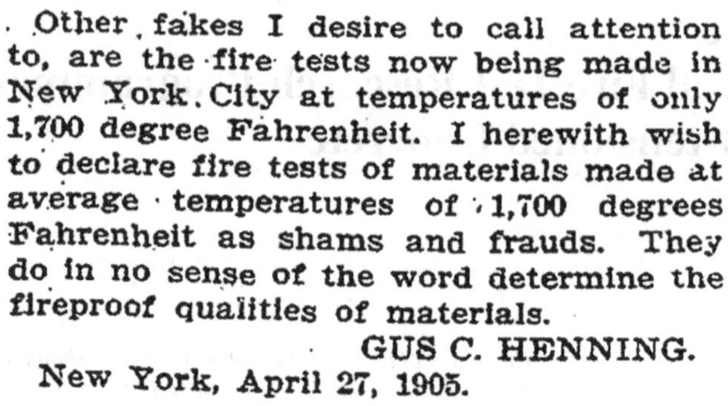

. Other fakes I desire to call attention
to, are the fire tests now being made in
New York City at temperatures of only
1,700 degree Fahrenheit. I herewith wish
to declare fire tests of materials made at
average temperatures of 1,700 degrees
Fahrenheit as shams and frauds. They
do in no sense of the word determine the
fireproof qualities of materials.

GUS C. HENNING.
New York, April 27, 1905.

Fig. 3.1 Gus Henning penned the above statement in the May 2nd 1905 New York Times. His reference to 1700 F (927 °C—the 1 h mark used in the standard fire today) was in relation that real fires would have larger temperatures and that materials would behave differently under more severe heating

under fire (for a complete historical account, see Gales et al. 2012 and Babrauskas and Williamson 1978).

The original requirements proposed by Himmelwright speak to good and best testing practice for structural systems. Over 100 years later, these practices are still being considered as relevant and acceptable methods for engineers to represent reality of the fire safety in our infrastructure.

That is where a potential problem lies with modern PT concrete structures, particularly those of highly optimized and complicated unbonded tendon configurations. This chapter expands on the previous efforts of the paper, *Unbonded Post Tensioned Concrete in Fire: A Review of Data from Furnace Tests and Real Fires* (Gales et al. 2011a). The intent of this chapter is to highlight the current state of knowledge with respect to the fire performance of PT concrete structures and to emphasize that traditional experimentation is incapable of realistically assessing real structural behavior in the event of a real fire or providing design guidance for such event, particularly of the unbonded system. This information serves as basis for the experimental program described in Chap. 4.

3.2 Furnace Tests of Unbonded Post-tensioned Concrete Members

There are at least 46 tests that can currently be considered for discussion on unbonded post tensioned (UPT) concrete members in fire. When the aforementioned paper was written (Gales et al. 2011a, b, c), there were only 27 tests to consider. Including these new tests, the majority of the conclusions previously compiled still hold true, but will be re-examined in the following chapter. The details of all tests can be found in the Appendix. This chapter provides a short description of the test data from UPT concrete members with their most profound and significant findings. Highlighted in this chapter are the deficiencies in current fire testing. One common occurrence is that when tests are made to move away from the standard fire, they do not follow 'consistent crudeness'. This term refers to the fact that simply changing one aspect of a test to deviate from the standard fire test is not necessarily defensible, and that all aspects of the test must have a comparable level of 'crudeness' for credibility (see Gales et al. 2012 for further discussion).

3.2.1   Legacy Testing

While dozens of fire tests on reinforced and prestressed concrete slabs and beams had been performed prior to 1980, only six tests have reported on flexural elements incorporating unbonded and stressed prestressing steel reinforcement. None of these tests can be considered as representative of current construction materials or techniques (Jimenez 2009), although they do provide some relevant information for contemporary critical commentary on restraint, steel mesh use, and pre-conditioning.

Interestingly, the earliest test ever conducted on UPT elements was among the most realistic. This test, led by the Fire Prevention Research Institute of California (FPRI), considered restrained and loaded ASTM E119 standard fire exposed concrete slab assemblies. The test was on a two-way UPT concrete beam-slab assembly. They were attempting to represent the two-dimensional response of the structure (Troxell 1959). The implication of this test is that it represented one of the earliest endeavors to consider the effects of structural stiffness, or restraint, on a building assembly in testing. This incorporation of restraint was 10 years ahead of common practice (see the compilation of the 1967 fire standards ASTM Symposium which debated the merits and pitfalls of testing building assemblies under restraint conditions). The test series was performed in the late 1950s, just as post-tensioning technology began to become common in concrete construction. Figure 3.2 illustrates details of the tested assembly.

Restraint for the beam-slab assembly was accomplished by fixing the test assemblies within a steel restraining frame along its entire perimeter. The purpose of this, according to that study's author, was "*to restrain it against thermal deformations and to simulate the conditions in a building.*" Neither the rotational nor lateral stiffness of the restraining frame were stated. Without the frame's stiffness, the resemblance of the test conditions to a real building is questionable. The test was stopped at approximately 4 h due to heat transmission criteria being exceeded. Spalling was observed on the beams but not the soffit. Longitudinal 'splitting' cracks about 1 mm in thickness along the cables were also observed (see Fig. 3.3 for an example of such a crack). The unexposed surface of the assembly developed cracks 6 mm wide adjacent to the beams due to thermal/flexural deformations.

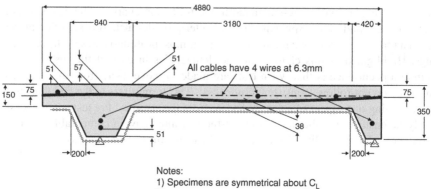

Notes:
1) Specimens are symmetrical about C_L
2) ∞∞∞ represents area exposed to fire
3) Bonded reinforcement not shown
4) All dimensions are in mm

Fig. 3.2 FPRI slab and beam test (Troxell 1959)

Fig. 3.3 An example of longitudinal cracking, taken from the slab fire tests described in Chap. 4. *Note*: the crack has been exaggerated through marking, its true thickness was approximate as 1 mm

In 1964, the Fire Research Laboratory of the Portland Cement Association (PCA), Illinois, conducted two standard fire tests of UPT beams with T-shaped cross sections (Gustaferro et al. 1971; PCI 1972; Gustaferro 1973). In this test series, the details regarding 'mesh' use and the difference between bonded and unbonded post-tensioning is of particular importance. Both UPT concrete beams included a light steel mesh placed at a mid-depth of the prestressing steel and the concrete cover, 25 mm. The mesh covered the entire fire-exposed perimeter of the beams. This is significant because the presence of a mesh may mitigate, and even arrest, cover spalling (Khoury and Anderberg 2000). The original PCA test report seems to ignore the abilities of mesh to reduce spalling, stating that "*the use of mesh within concrete cover is unnecessary*" (Gustaferro et al. 1971). Unfortunately, this opinion has since been embraced; the use of mesh in UPT structures is not common practice. Another commonly adopted conclusion from the PCA report is that beams with unbonded PT reinforcement have roughly the same fire endurance as their bonded counterparts (Gustaferro et al. 1971). This conclusion was made concerning isolated, simply supported, non-continuous beams with uniform heating. This is not at all how UPT structures are implemented in real building construction and therefore the wider applicability of this conclusion for real unbonded PT structures under localized heating is questionable. These tests were also not representative of the actual size PT beams are made to, and it has been shown by experiment that the longer the unbonded PT tendon, and the shorter the length over which it is heated, the more likely the tendon is to rupture prematurely in fire (Gales et al. 2011b, c).

In 1967, Underwriters Laboratory (UL) conducted a single standard fire test of a two-way restrained unbonded PT slab panel cast from lightweight concrete (UL 1968; PCI 1972; Gustaferro 1973). The test restraint was performed similarly to that by the FPRI. Figure 3.4 shows the tested assembly.

This test's significance is to highlight the act of preconditioning. Preconditioning is the act of artificially drying out the concrete prior to testing. By the 1960s it was well accepted that concrete spalling in fire was influenced by high moisture condi-

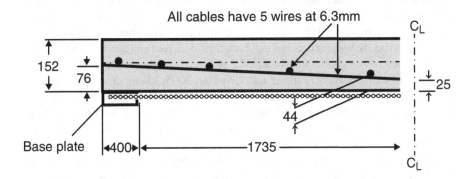

All cables have 5 wires at 6.3mm

Notes:
1) Specimens are symmetrical about C_L
2) **** represents are exposed to fire
3) Bonded reinforcement not shown
4) All dimensions are in mm

Fig. 3.4 Underwriters Laboratory slab (after UL 1968)

tions, in fact this behavior can be traced back to the 1850s (see Barrett 1854). Those who support preconditioning base their argument on a series of reinforced concrete slab fire tests conducted 70 years ago (Menzel 1943). Menzel performed a series of standard fire tests on preconditioned concrete assemblies. Some of those concrete assemblies underwent up to 9 h of standard fire exposure, an abnormally long time for a standard fire test. Several concrete slabs which had not been preconditioned were preserved for future monitoring, and their moisture content was assessed over time. After 12 years, they showed a similar moisture content to their preconditioned counterparts. Some North American researchers would argue that pre-conditioning to specific low moisture, as done so in these tests, is representative of in-service levels in real buildings. However, determining a realistic moisture content representative of real construction is quite challenging. Furthermore, since moisture contents in real concrete buildings have not been monitored to the awareness of the authors there is no way to clarify the issue and determine applicability of the practice. The UL test series preconditioned their PT slabs to excessive extremes, allowing the fire test to last for over 4 h. Preconditioning was performed at an elevated temperature of 49 °C and a relative humidity (RH) of approximately 20 % for 7.5 months prior to testing. This preconditioning makes the likelihood of cover spalling comparatively low. The RH of the slab at the time of testing was extremely low, about 1.5 % moisture by mass at the deepest sections. No spalling or cracking was observed on the exposed face during the fire test.

3.2.2 Modern Testing

Contemporary researchers have begun to question the wisdom of relying on standard fire test results from the pre-1980s to guarantee adequate fire resistance for UPT buildings. In particular, a 1983 study in Belgium (Van Heberghen and Van Damme 1983) (described below) recognized that both rotational restraint and tendon continuity into unheated bays could impact an UPT structure's response to fire. Recent testing in China and Germany (Falkner and Gerritzen 2005; Zheng et al. 2010; Aimin et al. 2013) have continued in this direction, however, they have failed to account for true axial, vertical and rotational restraints found in real unbonded PT structures, and have not used representative span-to-depth ratios typical of modern construction. As a result, these tests do not follow a consistent crudeness framework. More recently, tests have been performed in the UK (Bailey and Ellobody 2009a, b, c). In these cases the researchers reverted to performing single element tests of simply-supported, one-way spanning members with unrealistically small span-to-depth ratios and not accounting for continuity or realistic restraint.

Van Heberghen and Van Damme (1983) conducted an extensive series of non-standard structural fire tests on eight one-way continuous unrestrained UPT concrete slab strips. These tests appear to be the first fire tests ever to simulate rotational restraint through continuity at internal supports; however, they were conducted neglecting axial restraint. The tests attempted to rationally consider the possible

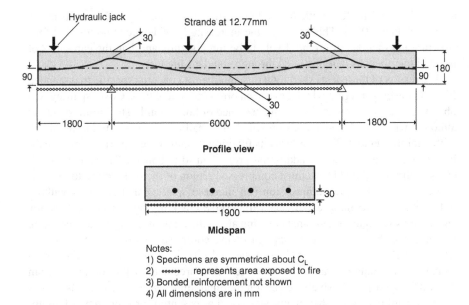

Fig. 3.5 Van Heberghen and Van Damme's slab (after Van Heberghen and Van Damme 1983)

influence of important parameters including: cover spalling, load ratio, and if present, the amount of "secondary" bonded mild steel reinforcement. Details of a typical specimen tested by Van Heberghen and Van Damme (1983) are shown in Fig. 3.5. Also shown is a schematic of the test setup, support and loading conditions, and length of fire exposure along the soffit.

The eight one-way slab strips were heated over the central span and one cantilever, giving a heated length ratio of approximately 80 %. The slabs were loaded during testing using four hydraulic jacks to produce "*a condition of zero rotation at the supports.*" Exactly how and why this loading condition was chosen is unclear. As a consequence of this rather unusual support condition, the applied load, and hence the bending moments at the critical sections, varied throughout the tests. It is unlikely that this was any more representative of full-structure response in a real fire than a restrained standard fire test on an isolated structural elements with constant vertical load. The initial load was chosen to simulate a comparatively high uniformly distributed superimposed live load. In the paper describing these tests, those authors comment that "*the importance of the initial bending moments was of minor influence on the ultimate fire resistance.*" It is not clear exactly what was meant by this, but it appears from the data presented, the evolution of bending moments during the fire exposure was far more important in governing the slabs' collapse than the initial loading condition. This is of considerable interest. The initial load ratio is universally assumed to be of central importance to the structural fire resistance of a flexural assembly in a standard fire test—to the extent that larger fire endurances are assigned to assemblies with lower load ratios for some types of construction (see Buchanan 2001). This confirms the increasingly widespread notion that full-structure interactions in fire are likely to be more significant than the capacities of isolated members as demonstrated through standard fire tests.

The results of these tests shed light on a number of concerns specific to unbonded PT concrete construction which have been observed in real fires. Cover spalling was observed in all tests regardless of the presence of bonded mild steel reinforcement; although when bonded steel was provided, the spalling was restricted to the depth of that reinforcement. This confirms its function in some cases to arrest cover spalling, providing a measure of additional protection to unbonded PT tendons. In all cases spalling initiated in the most compressed region of the soffit, close to the supports, showing that pre-compression can increase spalling in fire. Slabs without mild steel reinforcement experienced transverse cracking, both over the supports and at mid span, and longitudinal cracking along the tendons. Longitudinal cracking was noted to be due to a combination of pre-compressive stress combined with lateral tensile stresses generated from thermal effects during the fire. Slabs with increasing amounts of transverse mild steel reinforcement at the top and bottom faces displayed progressively fewer longitudinal splitting cracks. For this reason transverse mild steel reinforcement was noted as essential for minimizing longitudinal cracking. Subsequent researchers (Bailey and Ellobody 2009a; Zheng and Hou 2008) have also noted the importance of longitudinal splitting cracks on the fire endurance of unbonded PT structures. Some current design guidelines fail to properly address this issue.

Prestressing steel tendon rupture was observed in all eight tests. Ruptures were attributed to localized heating of the strands from exposure due to a combination of cover spalling, splitting along the tendons, and transverse cracking at mid span. The tendon rupture observed in this study is of particular interest for three reasons: first, these tendons were longer than those used in all previous testing; second, they were subjected to fire only over a portion of their length; and third, they were locally heated as a consequence of cracking and spalling as would occur in a real building fire. This shows that standard fire tests on short unbonded PT tendon lengths over single spans fully heated are incapable of rationally simulating the structural conditions in a real unbonded PT building. There is little doubt that traditional standard fire tests are un-conservative for evaluating the risk of premature tendon rupture. The results of the Belgian study lead to suggested revised minimum concrete cover depths, and minimum amounts of mild steel reinforcement in unbonded PT flat plate slabs and in support regions to prevent excessive flexural cracking. Current EN 1992-1-2 (2004) requirements appear to have been influenced by these recommendations (Kelly and Purkiss 2008). Other design guidance writing bodies (e.g. IBC 2012) appear to have largely ignored these recommendations (see Table 2.1 for comparative cover requirements between different countries), despite additional evidence from real fires (see Sect. 3.4).

Continuing in the direction of the Belgian study, in 2004 Litang et al. reported on three one-way continuous, unrestrained UPT slab strips exposed to an ISO 834 fire

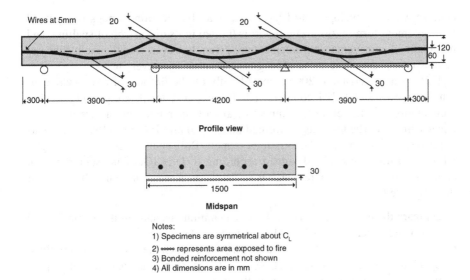

Profile view

Midspan

Notes:
1) Specimens are symmetrical about C$_L$
2) ∞∞∞ represents area exposed to fire
3) Bonded reinforcement not shown
4) All dimensions are in mm

Fig. 3.6 Litang et al.'s continuous slab (after Litang et al. 2004)

(Fig. 3.6). That research was originally written in Chinese and recently translated into English and republished (Aimin et al. 2013) with the addition of a fourth test. Each slab had three spans. The slabs were heated in their edge and middle spans. A brick wall within the furnace acted as a central support. The slabs were supported on one fixed (flat) support and three rollers. This prevented the slab from being restrained axially and vertically. Support reactions, relative stress loss, tendon temperature and deflection were all monitored during testing.

The prestressing force was not included in the English version of the study. The Chinese paper (Litang et al. 2004) indicates that study utilized un-representative initial stress levels prior to losses of approximately 40 % of the ultimate stress. Contemporary design guidance specifies about 70 % stress state (see Taranath 2010). Higher tendon loads, as in real construction, can lead to more pronounced relaxation effects due to accelerating plastic damage by temperature and load dependent creep. The higher loading also induces more precompression that can potentially induce spalling. Somehow, with only 30 mm of concrete cover, measured tendon temperatures remained below 350 °C even after 90 min of exposure the standard fire over a heated length ratio greater than 60 %. Tendon rupture was only confirmed in one out of the four tests. Discrete spalling was also found in the location of the rupture. The temperatures measured in the vicinity of the fracture were only 125 °C. These low temperatures and the low stress levels likely helped in preventing rupture in the three other tests. Deflections of all spans, with the exception of one test, indicated thermal bowing resulting in downward deflection, and an increase in tendon stress, rather than relaxation, was observed during testing. Support reactions indicated fluctuations up to 12 kN for all supports in all tests. Of principal concern was the development of large cracks at the location of negative bonded reinforcement termination. The report is limited in its discussion of spalling, and no details of transverse reinforcement are given, however, useful conclusions can be drawn from the observation of longitudinal cracking in every test.

Falkner and Gerritzen (2005) report briefly on the fire test of a one-way continuous, unrestrained UPT concrete slab. Limited test details are provided within their report, and therefore the impact of any of their findings on fire testing is equally limited. The test was performed under a 90 min ISO 834 (1999) fire heating two spans over a central continuous support. Both spans were 4.5 m in length, with a 200 mm thick slab designed according to DIN (German) standards. An equivalent reinforced concrete slab was also considered. The main objectives of the test were to:

- compare the deflection behavior between continuous, and similarly constructed reinforced concrete slabs, and UPT concrete slabs;
- provide numerical data for modelling purposes, which the researchers simplified by adding 3 kg/m^3 of polypropylene fibres to the fresh concrete mix in attempt to prevent explosive concrete cover spalling; and
- investigate the residual capacity of the slabs after fire exposure.

Falkner and Gerritzen (2005) found that the UPT slab displayed substantially less deformation than the reinforced concrete slab during testing (20 mm compared to 50 mm at test end). No spalling was reported but there was some distributed cracking along the exposed soffit of the slab. Falkner and Gerritzen hypothesized that this cracking was likely due to internal restraint stresses generated during the cooling phase of the test. Finally, almost full recovery of deformation for the PT slab was observed after cooling, and the final residual capacity was only decrease by 10 % when compared to a tested ambient slab. No repeat tests were performed, and little detail on tendon drape or cover are provided. While the data found through these tests could be of value for understanding how UPT structures behave in fire, whether or not the tests are representative of real building behavior cannot be ascertained.

Zheng and Hou performed two experimental programs. Their first series of one way tests as summarized in the Appendix. They continued their experimental program with a series of continuous slab tests, publishing first in 2007 in Chinese. The results of their study were then published in English in 2010 (Zheng et al. 2010). These new tests greatly resembled and were most likely inspired by those of Litang et al. (2004) and of Falkner and Gerritzen (2005) described above. Nine one-way continuous unrestrained UPT concrete slabs, with two 2.5 m spans each (Fig. 3.7), were tested. The slabs were lifted into a modified furnace, with a similar configuration to that presented by Litang et al. (2004), and placed onto three supports with two rollers and one flat plate support. Both spans were heated over 85 % of their total length with a target time temperature curve of ISO 834 (1999). With reference to their 2007 paper, their tested heating curve was not consistent with the ISO 834

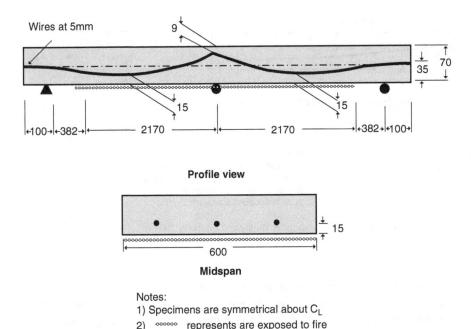

Profile view

Midspan

Notes:
1) Specimens are symmetrical about C_L
2) ∞∞∞∞ represents are exposed to fire
3) Bonded reinforcement not shown
4) All dimensions are in mm

Fig. 3.7 Zheng et al.'s continuous slab (after Zheng et al. 2007)

standard fire. This was not the case in the 2010 paper, where the figures showed the tested heating curve followed the standard fire with almost exact accuracy.

Discussion in the 2010 paper focuses on concrete cover spalling. The authors concluded that spalling took place between temperatures 200–500 °C and that the strength of concrete had a significant impact on spalling since concrete permeability decreases with higher compressive strengths. These insights support previous work by other researchers (see for instance Hertz (2003) and Dwaikat (2009)). All slabs exhibited similar deflection trends as those in Litang et al.'s (2004) tests. Support reactions were measured at unexposed slab ends with load cells. The load cells showed an initial decrease in load, followed by a rapid rise in load during cooling, although in all cases the reactions increased by less than 1 kN. Zheng hypothesized that this redistribution of force was due to changes in the deflected shape of the slab throughout the test. Significant tendon stress relaxation was observed. While an effort was made in these tests and others (Aimin et al. 2013; Falkner and Gerritzen 2005) to simulate a realistic continuous slab system, it is unlikely that secondary load balancing reactions would be representative given that these structural frames neglect axial, vertical and rotational restraint, all of which would be observed in a real structure with continuity over multiple supports.

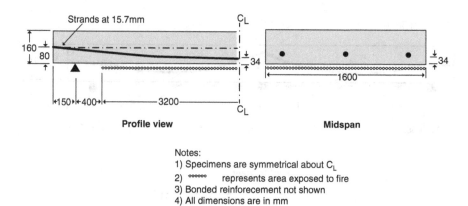

Fig. 3.8 Bailey and Ellobody's slab (after Bailey and Ellobody 2009a)

At the University of Manchester, Bailey and Ellobody (2009a) reported results of four standard fire tests on one-way spanning UPT concrete floor slabs. Lateral restraint was either provided or left unrestrained and the type of aggregate was varied between limestone and Thames gravel. These different aggregates were used to investigate the effect of varying thermal expansion on structural behavior. They also compare their results to the performance of bonded PT concrete slabs. Details of the UPT specimens are given in Fig. 3.8. The slabs were exposed to the BS EN 1991-1-2 (CEN 2004) fire until collapse in the case of the first test and in the three subsequent tests until critical temperature of prestressing steel (350 °C) was reached, to avoid furnace damage.

Two of the four UPT slabs were restrained longitudinally by two steel beams, but full details of the restraining mechanism are not available. Since the stiffness of the restraining frame is not known it is difficult to draw comparisons to real structures. None of the slabs experienced 'major' spalling during fire testing, although some localized spalling was visible. While all slabs developed longitudinal cracks, the unrestrained slab with limestone aggregate developed them on its unexposed surface after 20 min of fire exposure when the maximum recorded tendon temperature was only 108 °C. This slab collapsed into the furnace after 178 min. The collapse was apparently due to tendon rupture at the location of a major transverse flexural crack near mid span. The maximum tendon temperature prior to failure was 492 °C. The Thames gravel aggregate slabs performed similarly to the limestone aggregate slabs, except they experienced greater thermal expansion upon heating. Bailey and Ellobody (2009a) noted evidence of compressive membrane action and horizontal shear cracking during their restrained tests. They reiterate Van Heberghen and Van Damme's (1983) recommendation for transverse mild steel reinforcement to prevent longitudinal splitting cracking at the location of the tendons.

3.3   Furnace Tests of Bonded Post-tensioned Concrete Members

The following section comments on the influence of bonded versus unbonded construction. This discussion will provide better understanding of the structural mechanisms observed in UPT concrete fire tests and help explain how unbonded tendon continuity might affect floor performance in fire. Three modern test series are of interest: a single slab test by Kelly and Purkiss (2008), eight slabs tested by Bailey and Ellobody (2009a), and three slabs tested by Cement Concrete and Aggregates Australia (2010) (Fig. 3.9).

Kelly and Purkiss (2008) report an isolated test on banded tendon strands 8.5 m long, 3.6 m long, bonded in a 250 mm thick PT concrete slab. The slab was designed to have a 120 min fire rating; their test indicated excessive spalling and premature failure at 66 min. This isolated test has caused significant debate in the engineering community (Bailey and Ellobody 2009a). Bailey and Ellobody's view on these tests indicated that the moisture content was excessive at 4.6 % by weight which, with the use of gravel aggregate and supplied restraint, instigated spalling. Spalling remains a poorly understood behavior and more research is required to quantify whether this mechanism can be accounted for accurately in design. Nevertheless, these factors are all known to contribute to both bonded and unbonded PT construction.

Bailey and Ellobody (2009a) conducted a series of eight bonded PT concrete slab tests (both restrained and unrestrained) to complement their unbonded slab tests. The bonded slabs were designed to be identical to their unbonded counterparts for comparative reasons. The tests were conducted under a standard fire and limited by tendon temperature as an end test criteria. Bailey and Ellobody (2009a) describe the formation of the longitudinal cracks due to longitudinal tendon expansion, which relieves compressive forces in the concrete. The concrete is then subjected to tensile stresses caused by lateral expansion causing splitting at weaker sections (close to the tendons). This failure hypothesis is supported by their numerical modelling efforts. If this scenario is compared between the bonded and unbonded cases, crack formation mechanisms may be affected. In both the bonded

Fig. 3.9   A bonded post-tensioned concrete Duct with latex modified grout

and unbonded cases, this cracking behavior should arrest near the location where the prestressing steels are no longer exposed to high temperature. This would be due to the absence of tensile stresses from lateral expansion. However, since bonded tendons are cast with higher sized ducting than unbonded tendons, this further reduces the effective cross section of concrete that can carry tension. In bonded cases with no transverse reinforcement, this promotes tensile longitudinal cracking along the heated tendon length.

A series of tests were performed in 2010 by Cement, Concrete and Aggregates Australia (CCAA) in collaboration with the Centre for Environmental Safety and Risk Engineering. Three $1000 \times 150 \times 6000$ mm bonded unrestrained multi-strand PT slabs were tested under exposure to a real crib fire intended to simulate a standard fire for 70 min. The slabs were initially stressed to 1560 MPa using 12.7 mm diameter prestressing steel tendons. Significant spalling was noted during testing, and was attributed to the usage of gravel concrete. No longitudinal cracking was noted, but this could have been prevented by their use of transverse ties that were distributed through the length of the slabs. The only significant conclusion drawn from and analysis of test results was the confirmation that the addition of polypropylene fibres reduced the explosive nature of spalling.

The above tests illustrate many structural mechanisms which can also have consequences for all PT structures (spalling, cracking) and illustrate the importance on workmanship and construction for PT construction.

3.4 Real Fire Case Studies

In general, case studies of real fires in PT concrete buildings, particularly unbonded ones are rare. Gales et al. (2011a) compiles a listing of recorded and openly available case studies of UPT concrete buildings exposed to fire. Table 3.1 lists these in chronological sequence. Several important issues, including the consequences of tendon rupture to a structure, are identified through the compiled UPT concrete fire cases studies.

Partial collapse of an UPT slab exposed to fire for more than 5 h occurred in an 18 story building in Bangkok, Thailand in 1987 (Lukkunaprasit 1990). Each floor of the building had 4000 m² of two-way prestressed UPT flat plate construction, with interior bays of 80 m² each, and 4 m long cantilevers located at the end of each floor. The concrete slab was 200 mm thick and the cover to the prestressing steel tendons was 20–25 mm. Minimum mild steel reinforcement was provided according to the ACI provisions at that time. The fire started on the third floor and spread upward to the fifth floor. Widespread spalling occurred and exposed some of the tendons directly to the fire.

Eventually, some of the tendons ruptured and the cantilevers at the end of the fourth floor collapsed. This resulted in collapse of two supporting columns leading to the progressive collapse of several interior bays. It was estimated that 10–20 % of the tendons in the floor plate ruptured during the fire. Significantly, while bays in the

Table 3.1 Real fire case studies of unbonded post tensioned concrete buildings

Year	Publication	Location	Type of structure	Spalling (Y/N)	Fire	Progressive failure (Y/N)	Tendon Rupture (Y/N)
1965	Troxell (1965)	USA	UPT flat plate	Y	1.5 h peaking at about 1070°C	N	N
1987	Lukkunaprasit (1990)	Thailand	UPT flat plate	Y	Vertically travelling	Y	Y
1988	Barth and Aalami (1992)	USA	UPT flat plate	Y	Timber frame fire	Y	Y
1999	Sarkkinen (2006)	USA	UPT flat plate	Y	Timber frame fire	N	Y
2000	Post and Korman (2000)	USA	UPT flat plate	–	Vertically travelling	Y	Y
2000	Stern (2009)	Israel	UPT flat plate	–	Unspecified	N	Y

fire exposed region collapsed, adjacent bays did not fail even though the tendons were unbonded and continuous into the adjacent spans. Lukkunaprasit (1990) believed that tension membrane action of the slabs occurred with the tendons anchored at the edges of the collapsed bays by "kinks" in the tendon over the column lines. Large vertical displacements in the slabs are thought to have allowed a smaller tendon force to carry the weight of the slabs and the imposed loads by tensile membrane action. Tensile membrane action is widely recognized (Bailey et al. 2000) to occur in steel-concrete composite slabs in fire. On the basis of this fire, Lukkunaprasit suggested that engineers should supply 'unstressed' but bonded prestressing steel at the mid depth of UPT slabs to act as tensile membrane reinforcement and prevent collapse during a fire. Such measures have not yet been adopted into modern UPT design codes, nor has any subsequent research seriously considered this idea.

In 2000, a large uncontrolled fire occurred during construction of a 12 storey UPT concrete condominium building in Key Biscayne, Florida (Post and Korman 2000). The building contained unbonded and stressed prestressing steel tendons that were continuous across seven interior bays. The construction of the building had progressed to the 12th floor when a localized fire broke out on the second floor in an interior bay next to an internal shear wall. The fire spread to an adjacent bay where a pour-strip was located. A pour strip is an area of the slab where tendons are anchored during construction, it is left void until the building is nearing completion. The fire then spread vertically, causing visible fire damage up to the seventh floor across two interior bays (see Fig. 3.10).

According to the post-fire report by Post and Korman (2000), the engineers who examined the structure after the fire stated that "*heat caused the tendons at the pour strips to release tension.*" These engineers further noted that release of tendon tension "*triggered progressive failure of the post-tensioned slab well beyond the zone*

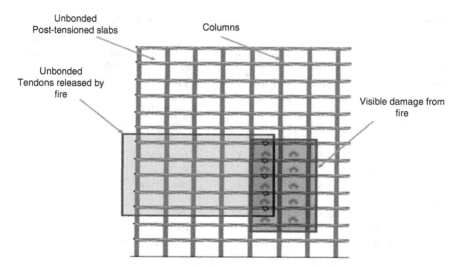

Fig. 3.10 Key Biscayne fire building profile

of visible damage." The result of this was that "*almost half the slabs on levels three to six, and possibly seven, lost integrity*". There was a loss of structural integrity across a 48 bays of the structure. The risk of progressive collapse was sufficiently high that no contractor would re-shore the floor slabs after the fire, and the entire building was demolished. Only the foundations were salvageable. The resulting economic losses from this fire were in the range of millions of dollars. This case study illustrates the potential consequences of localized tendon heating during fire and prestress loss across multiple bays. Brannigan and Corbett (2009) have also presented a case study of an UPT building that was subjected to a formwork fire during construction that led to collapse by a progressive mechanism, although this is less worrying given the unusual nature of the fire. The majority of other case studies emphasize significant occurrences of tendon rupture brought on by associated spalling and localized heating.

A case study presented by Stern (2002) described a UPT slab with spans up to 16 m exposed to a severe fire in Tel Aviv, Israel. Stern indicated that spalling of the concrete cover exposing the tendons occurred over a 300 m² area. Some of the tendons apparently ruptured and the mild steel was exposed during the fire. In this case, the UPT slab was reinstated by re-connecting and re-tensioning the ruptured tendons. The safety of this approach would have required considerable testing be performed to check for deterioration in post-heating residual mechanical properties of the prestressing steel (see Maclean et al. 2008; Roberston et al. 2013; Robertson et al. 2015).

The above case studies also highlight the absence of available knowledge in the community for assessing PT concrete structures post fire. Recently Robertson et al. (2015) presented a compilation of bonded PT concrete bridge fires. Three case studies were highlighted: one each from Europe, Canada and the USA. These case studies showed that concrete spalling may be a serious concern in a bridge fire due to the severe rapid heating associated to hydrocarbon fires, and addressed the severe economic costs associated to closing bridges post fire. The study identified a need to develop new technologies that can assess the residual tension and strength of the prestressing steel remaining after a fire.

3.5   Key Issues

The above sections raise a number of issues relevant to the fire-safe design of PT concrete structures, particularly of the unbonded state. UPT members appear to perform reasonably well in standard fire tests: in these scenarios they are typically able to comply with prescribed fire resistance ratings. However, given the large number of UPT structures in service, over 50,000 in the USA for example, and the relatively small number of real reported fires to compare this data to, it should not be construed as clear evidence of the fire safety of UPT structures. The following are considered the key issues identified in the available literary studies (see Appendix for more information):

3.5.1   Test Data

- With the exception of those to be discussed in Chap. 4, in the 44 described tests no realistic tests have ever been performed on multiple span continuous PT (unbonded or bonded) structures incorporating axial, vertical and rotational restraint.
- Most tests that demonstrate the fire safety of UPT slabs were performed prior to 1980 using construction materials that are now outdated, and in some cases using preconditioned specimens reducing the likelihood of cover spalling.
- Spalling was observed in at least 21 of the 44 described tests (45 %). It can be seen that limiting the moisture content to less than 3.0 % by mass, as is suggested in EN 1992-1-2 (CEN 2004), is clearly not sufficient to prevent spalling during fire in all cases. While the presence of non-prestressed mild steel reinforcement does not prevent spalling, it can help arrest spalling at the depth of the mild steel reinforcement. This can provide some protection to the prestressing steel tendons. More testing is needed to draw useful correlations between propensity for spalling and specimen parameters such as concrete strength, pre-compression levels, aggregate type, amount and location of mild steel reinforcement, load ratio, and concrete moisture content.
- Longitudinal splitting cracking along the tendons was observed in at least 22 of the 44 available tests (50 %). Current design guidance does not address this issue even though researchers have suggested that this contributes to and even causes failure of unbonded PT concrete structures in fire.

- Transverse cracks, in some cases as wide as 6 mm, were observed in at least 24 of the 44 available tests (55 %). Localized heating of tendons through these cracks could cause premature tendon rupture during fire.
- Tendon rupture during heating was observed in at least 12 of the 44 tests (27 %), and was particularly evident in tests with multiple spans and localized heating. Tendon rupture may be more likely in a real UPT building than in a furnace test of an isolated structural element or assembly.

3.5.2 Case Studies

- Some degree of concrete spalling occurred in all cases highlighted in Table 3.1. Localized spalling should be considered likely in all UPT buildings exposed to severe fires when preventative measures are not taken. It is therefore difficult to justify design for fire safety of UPT systems purely on the basis of minimum concrete cover to the tendons, as specified in Table 2.1, as it is based on the assumption that the concrete cover will remain in place during fire.
- Tendon rupture or release of prestress occurred in two thirds of the UPT concrete case studies. This confirms that tendon rupture, or prestress loss by tendon relaxation, is likely to occur in a UPT building in a real fire. This has led to both partial and progressive failure of real UPT buildings during fire.
- UPT structures exposed to real fires are almost certain to experience non-uniform and localized heating, which will expose the tendons to heating over a portion of their total length.

3.6 Summary

The goal of fire testing structural materials should be to provide a practical design for life safety and property protection. The earliest set of requirements for meeting these objectives were noted by Abraham Himmelwright (1898) in a precursor to the ASTM E119 standard fire test. The conditions Himmelwright specified are: (*1*) *that tested materials should be identical to commercially available ones*; (*2*) *that tested materials should be constructed as in practice*; *and* (*3*) *that the fire tests should be conducted in a scientific manner.* The UPT concrete standard fire tests discussed above raise research concerns with regard to current capabilities in meeting these century old crucial objectives. It is widely acknowledged that standard fire testing is unrealistic for most real structures (Bisby et al. 2013), however, even when researchers have deviated from the standard fire test they typically have not followed a consistent crudeness framework. Credible concerns for meeting the goals of structural fire safety in UPT concrete buildings remains a healthy debate.

It is important to expose the unrealistic nature of standard fire testing for UPT concrete beams and slabs. Standard fire testing is fundamentally incapable of realistically simulating several important and interrelated behaviors that could be expected and that have been directly observed in real UPT buildings during real fires. These behaviors can lead to premature tendon rupture and progressive failure of the entire building. Localized heating of unbonded and stressed prestressing steel tendons may occur during localized or travelling fires in multi-bay structures due to a combination of:

- single bay, localized, or travelling fires in multi-bay structures;
- inconsistent and inadequate concrete cover for draped tendons;
- spalling of the concrete cover; and/or
- longitudinal and/or transverse cracking of the cover.

Localized heating of an unbonded and stressed prestressing steel tendon is likely to rapidly lead to rupture, as has been seen in many tests and real fires. The risk and consequences of localized heating from spalling can be mitigated through the use of bonded mild steel reinforcement, although this issue is still neglected in some design standards. Bonded mild steel reinforcement can limit the depth of spalling, promote a finer and more evenly distributed cracking pattern, and permit alternative load carrying mechanisms. Research is needed to define the appropriate minimum amount and placement of bonded reinforcement to sufficiently aid in preventing collapse during fire.

Because full-scale fire tests on actual or model UPT buildings are unlikely to occur in the foreseeable future research is currently restricted to using computational analysis tools (e.g. Bailey and Ellobody 2009a, b, c) to study their response to fire. In general, these tools have not been validated against real fires in realistic UPT concrete structures and their ability to accurately model various aspects of UPT concrete slabs at elevated temperature is therefore unreliable. A detailed experimental and computational examination of the potential consequences of localized heating on UPT tendons is needed in order to eventually develop a rational understanding of the performance of real UPT buildings in real fires. The tests presented in Chap. 4 address this need.

References

Aimin, Y., Yuli, D., & Litang, G. (2013). Behaviour of unbonded prestressed continuous concrete slabs with the middle span and edge span subjected to fire in order. *Fire Safety Journal, 56*, 20–29.

ASTM. (2014). *Test method E119-14: standard methods of fire test of building construction and materials* (Rep. No. E119-14). West Conshohocken, PA: American Society for Testing and Materials.

Babrauskas, V., & Williamson, R. B. (1978). The historical basis of fire resistance testing – Part 1. *Fire Technology*, 304–315.

Bailey, C. G., White, D., & Moore, D. (2000). The tensile membrane action of unrestrained composite slabs simulated under fire conditions. *Engineering Structures, 22*, 1583–1595.

Bailey, C. G., & Ellobody, E. (2009a). Comparison of unbonded and bonded post-tensioned concrete slabs under fire conditions. *The Structural Engineer, 87*(19), 23–31.

Bailey, C. G., & Ellobody, E. (2009b). Fire tests on unbonded post-tensioned one-way concrete slabs. *Magazine of Concrete Research, 61*(1), 67–76.

Bailey, C. G., & Ellobody, E. (2009c). Whole-building behaviour of bonded post-tensioned concrete floor plates exposed to fire. *Engineering Structures, 31*, 1800–1810.

Barret, J. (1854). On the construction of fire proof buildings. *Proceedings of the Institution of Civil Engineers, 883*, 244–272.

Barth, F., & Aalami, B. (1992). *Controlled demolition of an unbonded post-tensioned concrete slab* (PTI special report). Post-tensioning Insititute, Pheonix, Arizona, 34 pp.

Bisby, L., Gales, J., & Maluk, C. (2013). A contemporary review of large-scale non standard structural fire testing (circa 1980–present). *Fire Science Reviews, 2*(1).

Brannigan, F., & Corbett, G. (2009). *Building construction for the fire service* (4th ed.). Sudbury, MA: Jones and Bartlett.

Buchanan, A. H. (2001). *Structural design for fire safety*. New York, NY: Wiley.

CCAA. (2010). *Fire safety of concrete buildings*. Cement Concrete and Aggregates Australia – CCAA T61, 37 pp.

CEN. (2004). *Eurocode 2: Design of concrete structures, Parts 1–2: General rules-structural fire design, EN 1992-1-2*. Brussels: European Committee for standardization.

Dwaikat, M. B., & Kodur, V. (2009). Hydrothermal model for predicting fire-induced spalling in concrete structural systems. *Fire Safety Journal, 44*(3), 425–434.

Falkner, H., & Gerritzen, D. (2005). Reuse of slabs prestressed with unbonded tendons after fire exposure. *IABSE Symposium on Structures and Extreme Events, 90*, 23–30.

Gales, J., Bisby, L., & Gillie, M. (2011a). Unbonded post tensioned concrete in fire: A review of data from furnace tests and real fires. *Fire Safety Journal, 46*(4), 151–163.

Gales, J., Bisby, L., & Gillie, M. (2011b). Unbonded post tensioned concrete slabs in fire – Part I – Experimental response of unbonded tendons under transient localized heating. *Journal of Structural Fire Engineering, 2*(3), 139–154.

Gales, J., Bisby, L., & Gillie, M. (2011c). Unbonded post tensioned concrete slabs in fire – Part II – Modelling tendon response and the consequences of localized heating. *Journal of Structural Fire Engineering, 2*(3), 155–172.

Gales, J., Bisby, L., & Maluk, C. (2012). Structural fire testing – Where are we, how did we get here, and where are we going? In: *Proceedings of the 15th International Conference on Experimental Mechanics*, Porto, Portugal, 22 pp.

Gustaferro, A. H. (1973). Fire resistance of post-tensioned structures. *PCI Journal, 18*(2), 38–62.

Gustaferro, A. H., Abrams, M., & Salse, E. (1971). Fire resistance of prestressed concrete beams study C: Structural behaviour during fire tests. *PCA Research and Development Bulletin, 29*, 1–29.

Hertz, K. D. (2003). Limits of spalling of fire-exposed concrete. *Fire Safety Journal, 38*, 103–116.

Himmelwright, A. (1898). Fire proof construction. *The Polytechnic*, 167–175.

IBC. (2012). *International building code*. USA: International Code Council. Country Club Hills, Il.

ISO 834. (1999). *Fire resistance test – Elements of building construction*. Geneva: International Organization for Standardization.

Jimenez, G. A. (2009). Assessment and restoration of post-tensioned buildings – Parking ramp structures. In *Proceedings of the 2009 Structures Congress ASCE* (pp. 1954–1963).

Kelly, F., & Purkiss, J. (2008). Reinforced concrete structures in fire: A review of current rules. *The Structural Engineer, 86*(19), 33–39.

Khoury, G. A., & Anderberg, Y. (2000). *Concrete spalling review, fire safety design* (Report submitted to Swedish National Road Administration).

Litang, G., Dong, Y., & Aimin, Y. (2004). Experimental investigation of the behaviour of continuous slabs of unbonded prestressed concrete with End span under fire. *Journal of Building Structures, 25*(2), 118–123 (in Chinese).

Lukkunaprasit, P. (1990). Unbonded post-tensioned concrete flat plates under 5-hours of fire. In *11th FIP congress in Hamburg Germany*, 61–64.

MacLean, K., Bisby, LA., & MacDougall, CC. (2008). Post-fire Assessment of Unbonded Post-tensioned Slabs: Strand Deterioration and Prestress Loss. *ACI-SP 255: Designing Concrete Structures for Fire Safety, American Concrete Institute*, 10 pp.

Menzel, C. (1943). Tests of the fire resistance and thermal properties of solid concrete slabs and their significance. In *Proceedings of the ASTM 43*, 1099–1153.

Post, N., & Korman, R. (2000). Implosion spares foundations. *Engineering News Record, 12*, 12–13.

Roberston, L., Dudorova, Z., Gales, J., Stratford, T., & Blackford, J. (19–20 April 2013). *Microstructural and mechanical characterization of post-tensioning tendons following elevated temperature exposure*. Applications of Structural Engineering Conference, Prague, CZ, pp. 474–479.

Robertson, L., Gales, J., & Stratford, T. (2015). Post-fire investigations of prestressed concrete structures. In: *Proceedings of the Fifth International Workshop on Performance, Protection & Strengthening of Structures under Extreme Loading*, Michigan, USA, pp. 747–754.

Sarkkinen, D. (2006, June). Fire damaged post-tensioned slabs. *STRUCTURE Magazine*, 32–34.

Stern, I. (2002). *Restoration of long span plate post-tensioned with unbonded tendons – After fire*. Lecture given at the 2002 fib Congress, Osaka, Japan. (online) Available at http://www.yde.co.il/. Accessed Jan 1, 2015.

Taranath, B. S. (2010). *Reinforced concrete design of tall buildings*. Boca Raton: CRC.

Troxell, G. E. (1959). Fire resistance of a prestressed concrete floor panel. *Journal of the American Concrete Institute, 56*(8), 97–105.

Troxell, G. E. (1965). Prestressed lift slabs withstand fire. *ASCE Civil Engineer*, September, 64–66.

Underwriters Laboratories. (1968). Report on unbonded post-tensioned prestressed reinforced concrete flat plate floor with expanded shale aggregate. *PCI Journal, 4*, 45–56.

Van Heberghen, P., & Van Damme, M. (1983). *Fire resistance of post-tensioned continuous flat floor slabs with unbonded tendons* (FIP Notes), pp. 3–11.

Woolson, I. H. (1902, October 19). Making buildings safe: Fire proof materials and methods of construction. *New York Tribune*.

Woolson, I., & Miller, R. (1912). Fire tests of floors in the United States. In *International Association for Testing Materials 6th Congress*, New York.

Zheng, W., Hou, X. M., Shi, D. S., & Xu, M. X. (2010). Experimental study on concrete spalling in prestressed slabs subjected to fire. *Fire Safety Journal, 45*, 283–297.

Zheng, W., Hou, X., & Xu, M. (2007). Experiment and analysis on fire resistance of two-span unbonded prestressed concrete continuous slabs. *Journal of Building Structures, 28*(5), 13 (in Chinese).

Zheng, W., & Hou, X. (2008). Experiment and analysis on the mechanical behaviour of PC simply supported slabs subjected to fire. *Advances in Structural Engineering, 11*(1), 71–89.

Chapter 4
Localized Heating of Post-tensioned Concrete Slabs Research Program

If we attempt to develop the fire endurance of a construction system in actual buildings under fire conditions we would not obtain a single-valued answer, but rather we would have to measure a range of performance levels depending upon methods of structural framing existing in a single building as well as the methods of structural framing of any and all buildings into which the construction system under consideration could be incorporated....

—RW Bletzacker (1967)

4.1 Background

Performance-based design is the growing paradigm in contemporary structural engineering, and structural fire safety engineering is no exception. Advocates of performance-based methodologies seek to adopt sophisticated fire strategies tailored to individual building needs. In particular, these strategies are being applied to optimized reinforced concrete buildings, including post-tensioned (PT) structures. The current understanding of prestressing steel behavior based largely on outdated research that fails to properly account for material property changes at elevated temperatures. Furthermore, real fires in real PT concrete buildings have the potential to induce unique failure mechanisms that cannot be observed or accounted for using standard fire tests, as seen through case studies of real fires (see Chap. 3). Current modelling tools used to establish structural fire safety engineering strategies lack realistic experimental validation and verification, leading to the development of potentially unconservative performance-based strategies for PT concrete buildings. Indeed for all concrete buildings. In order to enable credible performance based design of PT concrete buildings current modelling capabilities need to be improved, through the analysis of densely instrumented experiments, which incorporate as many relevant structural properties (post-tensioning, continuity, restraint, realistic scale, unbonded reinforcement, etc.) of as-built PT construction as possible.

© The Author(s) 2016
J. Gales et al., *Structural Fire Performance of Contemporary Post-tensioned Concrete Construction*, SpringerBriefs in Fire,
DOI 10.1007/978-1-4939-3280-1_4

This need is partly addressed in the current chapter by presenting experiments on three 3-span continuous, restrained PT concrete slabs (slab strips). The slabs were put under sustained service loading and exposed to severe localized heating using radiant heaters. This chapter considers an extensive recent experimental program conducted for this book over the years 2011–2013. This experimental program compliments those discussed in previous chapters. Three novel large-scale tests on locally heated, continuous PT concrete slab strips are detailed. In real and contemporary multiple bay and continuous unbonded post tensioned (UPT) concrete slabs exposed to fire, structural actions may play significant (often interrelated) roles influencing the response of this construction. This response is studied and explained herein.

4.2 Motivation and Objectives

When real multi-bay continuous UPT concrete structures are exposed to fire, thermal deformations, restraint, reductions in concrete stiffness, tendon stress relaxation, floor continuity, spalling, slab splitting, etc., may all influence the structural response. To date, relatively little experimental investigation on these structural interactions in fire using realistic PT concrete slab construction configurations has been conducted. To guide the structural and fire engineering community towards the development of defensible, performance based and fire-safe designs for these structures, a series of non-standard structural fire experiments were performed on continuous and restrained PT concrete slab strips.

Earlier research indicated that localized heating caused by travelling fires and ceiling jets has a more severe effect on premature rupture of unbonded prestressing steel tendons than uniform fires (Gales et al. 2011a, b, c). Localized heating is also more likely to be the case in real fires. The global response of a continuous UPT concrete slab system to fire has not yet been considered. Little experimental data exists to help designers understand the fire performance of realistically constructed and continuous restrained UPT concrete slabs, despite the industry's rapid optimization trends for ambient design. The experiments presented in this chapter were designed to study the structural and thermal response of simplified continuous PT concrete slabs under highly localized heating while incorporating as many of the relevant structural behaviors as possible. Such tests represent a critical step in the development of a rational understanding of the fire performance of PT concrete structures, and therefore the ability to create realistic and credible modelling tools.

Three one-way continuous, restrained, monostrand PT concrete slabs were constructed and tested under localized heating. Two of the slabs had unbonded tendons, and the third had a bonded tendon. All slabs were precast and then connected to steel column restraints/supports prior to testing. These columns were designed to be representative of the axial-flexural restraining stiffness that would be present in a real PT concrete building. This resulted in three spans on four partially rigid and fixed connections. After installation, the prestressing tendons were post-tensioned and anchored. This allowed possible restraining mechanisms during heating to be quantified and investigated. All previous testing presented in literature has been unable to represent this condition as it has used either roller connections or flat plate frictional bearings at support locations (see Chap. 3). The columns used for this experimental series were instrumented with strain gauges which made it possible to monitor the restraining forces developed during testing. Deformations were monitored during testing with displacement transducers, and tendon stress levels with load cells. Each slab was statically loaded with lead weights during testing to simulate a representative in-service condition. These weights were used because they can remain stationary during small deflections and can be directly exposed to temperatures up to 300 °C. Localized heating was applied using an array of radiant heaters placed underneath the central span. Temperatures were monitored using thermocouples embedded in each slab as well as a thermal imaging camera to measure the soffit temperatures. The slabs were monitored for structural and thermal responses in both heating and cooling phases.

All slabs had the same overall dimensions, thickness, axis distance to the prestressing tendon, concrete type, steel reinforcement, age, and moisture condition at the time of testing. They were constructed to investigate the influence of heated length ratio on the performance of the tendons in a real structure, as well as to directly investigate possible differences between bonded and unbonded tendon configurations in concrete construction. The three slabs (A–C) are described as follows:

- **Slab A** was the base case slab. The prestressing tendon was installed as unbonded (greased and sheathed) and subjected to heating over its central span, at 17 % of its total length.
- **Slab B** was included to investigate the influence of bonded versus unbonded construction. The only difference from Slab A was that the prestressing tendon was installed as bonded (i.e. grouted within a plastic post-tensioning duct).
- **Slab C** was identical to Slab A with the exception that a custom-built disc spring was installed at its dead end anchorage. This decreased the stiffness of the unbonded tendon in order to simulate a longer unbonded tendon length and a shorter heated length ratio (less than about 10 % in this case).

Temperatures, tendon stress levels and stress relaxation, restraining forces, and deflections were all monitored during testing. Since very little data is available for the cooling phase response of concrete structures, all three slabs were monitored during cooling.

Selected details of the test specimens are provided in Table 4.1. Figure 4.1 gives a schematic of the slab, support, and heating systems. Figure 4.2 gives a schematic view of the slab and its instrumentation.

Table 4.1 Selected details of test specimens

Slab	Type	Precompression (MPa)[a]	Restraint conditions (lat/rotation)[b]	Bonded steel reinf.[c] (%)	$f_{C_{cube}}$ at 28 days (MPa)	Moisture content at testing (% by weight)	Aggregate type	Load ratio[d]
A	Unbonded	2.37	Y/Y	0.22/0	41	4.0	10 mm mixed gravel	0.42
B	Bonded	2.42		0.22/0	42	3.9		0.32
C	Unbonded	2.39			43	3.8		0.42

Slab	Type	Span/depth ratio	Longitudinal crack	Max. tendon temp (°C)[e]	Transverse crack	Spall	Tendon rupture?	End Point
A	Un-bonded	43	N	361	Y	N	N	Critical temperature
B	Bonded		Y	376	Y	Y	N	
C	Un-bonded		?[f]	432	Y	N	N	

[a]Precompression is the stress level at test start divided by the slab cross section
[b]Y yes, N no, P partial
[c]Top longitudinal/top transverse/bottom longitudinal/bottom transverse
[d]Load ratio is applied moment reaction to theoretical moment capacity at mid span
[e]Maximum temperature occurred after heating ceased to a small amount
[f]Longitudinal crack not measurable on surface of slab, but flaming parallel with the tendon was observed during the test

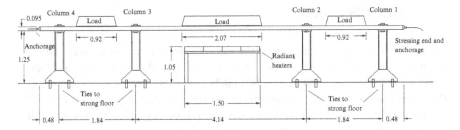

Fig. 4.1 Schematic showing the test set up and selected geometry of every slab (dimensions in m, instrumentation and insulation boards not shown)

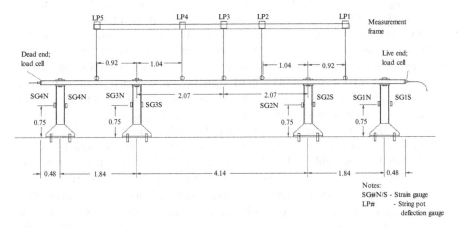

Fig. 4.2 Schematic showing the instrumentation set up and selected geometry (dimensions in m, insulation boards not shown, thermocouples are described in Fig. 4.11)

4.3 Experimental Overview

In Chaps. 2 and 3, it was demonstrated that most prior tests on PT concrete slabs have used materials and techniques that are too old to be representative, dimensions which cannot be considered typical of modern construction, and support conditions that do not resemble reality. The test explained in this chapter begin, for the first time, to explore these potentially important issues.

Each of the slabs was precast with eight bolt hole connection points and four lift hooks. After casting and curing (for a minimum of 6 months) they were lifted into place on top of four semi-rigid steel columns. A rigid connection was made between the slabs and the columns, and the slabs were then post-tensioned to representative in-service prestressing levels. An imposed service loading was then applied prior to heating.

4.3.1 Slab Dimensions

The principal goal in designing the slabs was to develop a representative, one-way spanning restrained continuous monostrand PT concrete slab. The slab spans therefore needed to be restrained vertically, axially, and rotationally at all connections. This is representative of the conditions for an interior span in a real building. The experimental programme called for the design of three bays with two short cantilevers at both ends. The minimum slab width-to-depth ratio of five was chosen, in accordance with CL 5.3.1 (4) of EN 1992-1-2 (CEN 2004) so as to be classified as a slab rather than a beam (five to one rule). Additionally, a minimum 1 h fire resistance rating was sought with respect to the overall thickness of the slab. The thickness and width of the slab were chosen accordingly, which resulted in a design slab width of 475 mm with a total depth (thickness) of 95 mm. A minimum span-to-depth ratio of 40 was necessary for the center span to be representative of modern PT concrete construction. The length of the center span was therefore defined as 4140 mm. The total length of the slab (8770 mm) was restricted by the maximum dimensions of the strong floor in the testing laboratory. Cantilevers outside the end supports accommodated bursting reinforcement in the prestressing anchorage zones. The slab dimensions are illustrated in Fig. 4.3. Concrete cover was chosen representing a nominal 1-h fire resistance rating according to EN 1992-1-2 (CEN 2004), with bonded mild steel reinforcing bars at 25 mm cover and tendon axis distances at 35 mm at mid span (and 35 mm to the top surface at supports).

The specified drape of the monostrand seven-wire 12.5 mm diameter Grade 1860 prestressing tendon followed a parabolic profile to provide balancing of the applied service load, as would be the case in design of a real PT concrete building. The drape is illustrated in Fig. 4.3. To ensure that Slab B would have the same axis distance after tensioning as Slabs A and C, the prestressing duct for this slab was positioned before casting to account for possible alignment changes during tensioning. The prestressing duct was supplied by GT Technologies™ (with 23 mm inner diameter), it is not shown in Fig. 4.3.

The slabs were required to have sufficient positive moment reinforcement running the length of the slab to satisfy the expected lifting stresses without cracking. All non-prestressed steel reinforcement was 8 mm diameter deformed Grade 500 reinforcement. Negative moment reinforcement was provided over all supports, as indicated in Fig. 4.3. The final slab configuration was verified for in-service stresses, flexural, shear and final expected camber. The design is representative of a strip of a one-way PT slab that would be expected in a real building, with load balancing effects typical of real PT slabs.

4.3.2 Supporting Column Design

A set of elastic steel column supports were used to support and restrain the slabs during testing. These were designed on the basis of their axial and flexural stiffness at ambient temperature. The columns were selected based on restraining deflection

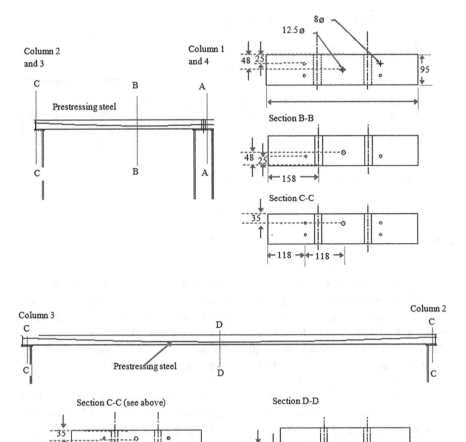

Fig. 4.3 Slab dimensions including axis distance covers (dimensions in mm)

compatibility. A representative concrete column in a real PT building might have dimensions of 450×450 mm, an internal steel reinforcement ratio of 0.05 %, a specified concrete compressive strength of 50 MPa, and a height of 2.6 m (see CPCI 2007; CSA 2004; CSTR 2005). An appropriate commercially available steel section was selected to nominally match these properties, a $203 \times 203 \times 60$ I-section made from Grade 275 steel. The column height was taken as 1.25 m for convenience during testing. While care was taken to select a representative column size for the testing rig, clearly not all buildings are constructed according to the above dimensions. Real column designs will depend on the expected applied loading, floor configurations, and architectural demands. This comment is especially important when considering the restraining forces induced during the tests, and will be discussed in detail. The concrete slab-to-steel column connection was made in accordance with EC3-1-8

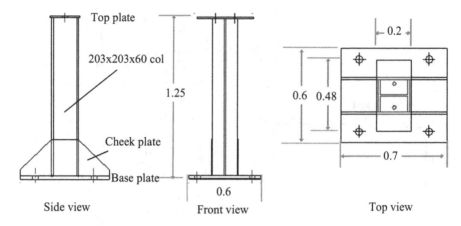

Fig. 4.4 Column support schematics (dimensions in m)

(CEN 2005). A steel top plate of dimensions $475 \times 203 \times 25$ mm, with two bolt holes cored at 20 mm diameter was fixed the column to allow a seat for the slab. The top plates were welded completely to the tops of the steel columns to minimize the flexibility of this connection. Column base plates of dimensions $700 \times 600 \times 25$ mm, each with four bolt holes cored for 36 mm diameter anchor bolts, were added to the bases of the columns. Two plates were added to increase the rigidity of the base plate detail. Figures 4.4 and 4.5 give schematics for the columns.

Each column was bolted and tied into a strong floor using four (Grade 275) 36 mm diameter bolts through two 610 mm steel channel sections (size $150 \times 75 \times 18$). Prior to installation of the columns on the strong floor, a detailed survey of the strong floor was necessary to ensure that the column tops would be level so that the slab would not be exposed to settling moments when placed on top of the columns. A calibration exercise to gauge the stiffness of the column was conducted through a load unload exercise prior to slab placement.

4.3.3 Post-tensioning Anchorage Zone Design

The anchorage zone reinforcement design followed guidance from BS 8110-1 (BS 1997) and the CPCI design manual (CPCI 2007). Bursting stresses were determined based on a 95×95 mm bearing plate (with prestressing anchors resting against the bearing plate). The design included reinforcement for spalling, bearing, and end stress accumulation at transfer during jacking. This design was meant to accommodate a 130 kN applied load on the surface of the concrete, whereas in the tests the prestressing steel in the slabs was tensioned to approximately 125 kN before losses. All reinforcement links in the bursting zone were made of 6 mm diameter mild steel reinforcement (tensile strength 610 ± 5 MPa). Coiled reinforcement with 3 mm

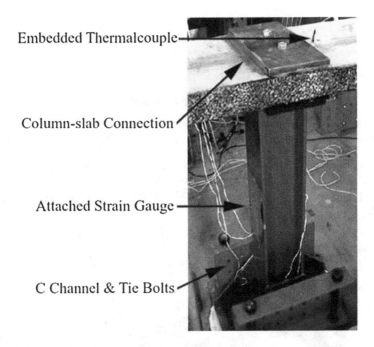

Fig. 4.5 Column support after connection to strong floor and slab

diameter and tensile strength of 355 MPa was used for confinement of the bursting zone. The assembled anchorages for both live and dead ends, including the load cell and barrel wedge anchors, are shown in Fig. 4.6. An adjusting screw was also installed in the dead end anchorage (Fig. 4.6) to help de-tensioning the unbonded prestressing steel after testing.

To simulate a smaller heated length ratio without exceeding the dimensions of the strong floor, an industrial spring mechanism was built and installed at the dead end anchorage for Slab C. The anchorage system (shown in Fig. 4.7) was constructed with 36 disk springs (200 mm outer diameter, 102 mm inner diameter, and thickness 4.5 mm with an uncompressed height of 12.5 mm) stacked in series and enclosed within a cylindrical steel shaft. The spring assembly was connected to a fixed bearing plate which rested against the end of the slab. A second 'free' plate acted on the top of the spring stack inside the sheath. A hole through the spring assembly allowed the tendon to pass through its center. Thus, when the prestressing tendon was tensioned the free bearing plate compressed the spring stack.

The compression of the spring stack is representative of a decreased effective stiffness of the unbonded tendon, thereby simulating the structural effects of a longer unbonded tendon length. A calibration exercise on the disc-spring anchorage (compressed under load) confirmed the added simulated tendon length as an additional 8.5 m, simulating a total tendon length of approximately 17.3 m as opposed to the true length of 8.8 m.

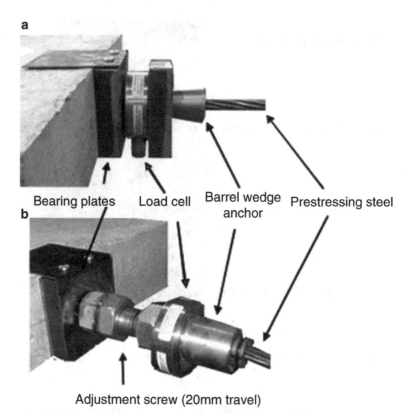

Fig. 4.6 (**a**) Stressing (live) end and (**b**) dead end

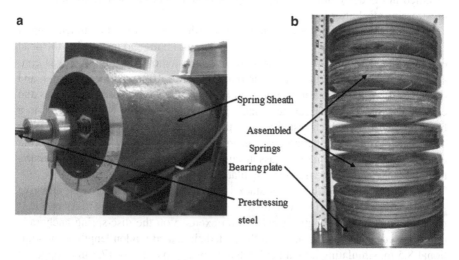

Fig. 4.7 (**a**) Full spring assembly bearing on slab end, (**b**) spring stack

4.3.4 Applied Loading

The fire test load ratio represents the applied loads (self-weight plus partitions and imposed loads) divided by the predicted capacity of the slab at ambient temperature. Considering the small number of slabs that could be cast and tested within the time and space limitations, no slabs were tested to failure at ambient temperature. For this reason, a theoretical analysis of the ambient temperature strength was conducted.

The capacity of a PT slab will be influenced by the loading configuration, span to depth ratio, unbonded tendon length, etc. (Allouche et al. 1998, 1999). When prestressing steel is unbonded, simple flexural strain compatibility cannot be relied on as is the case bonded tendons. The theoretical capacity for the base case (Slab A) was approximated as 7.3 kNm using guidance from CSA (2004) and Ghalleb (2013). Applied distributed loading for each span was provided using 70 individually weighed lead weights applied over the central half of each span (see Fig. 4.1). This was maintained constant throughout heating and gave a load ratio equal to 0.42 with respect to the theoretical capacity of Slab A at ambient temperature (assuming all material safety factors taken as unity). A load ratio of below 0.5 can be considered a likely service condition in the event of a fire (Buchanan 2001).

4.3.5 Heating Array

An array of four radiant heaters, each constructed of cast-iron bodies with ceramic combustion surfaces, was mounted on a horizontal steel frame in order to heat the slab. Figure 4.8 shows the heating assembly. Figure 4.9 shows the heating array in situ underneath a slab during testing. The purpose of attached ceramic fiber boards was ensure localized heating by preventing direct heating of adjacent portions of the slab and the supporting columns.

Propane and air were fed into mixer valves mounted on the back or each panel. Air was supplied by four electric blowers. A mass flow controller was used to regulate the gas flow upstream of the gas manifold, and the flow of propane was maintained at 1.4 g/s during all tests. The burners were lit with automatic sparkers. The radiant panel heater array had a total length of 1500 mm and width of 480 mm, with an intention to subject the slabs to uniform heating over their central spans, about 17 % of their total tendon length. Slab C, with additional spring-simulated tendon length, had a simulated heated length ratio intended to be 9–10 %.

Due to heat losses, the system used cannot apply as severe heating as would occur during a 'standard' ASTM E119 (ASTM 2014) or ISO 834 (ISO 1999) time-temperature curve; however, the intent of this experiment was not to develop certification with standard fire resistance ratings which have little bearing on reality. The intent of this experimental program was to investigate possible structural performance of PT structures under 'well-defined' heating and to generate thermal data for future numerical modelling techniques and design.

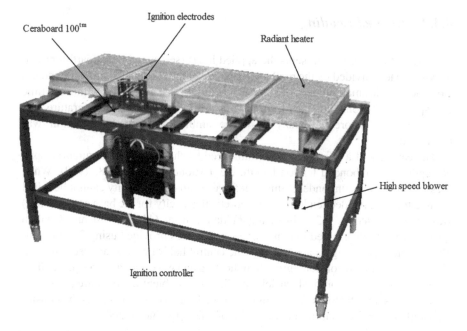

Fig. 4.8 Radiant panels and heater assembly (debris cage not shown)

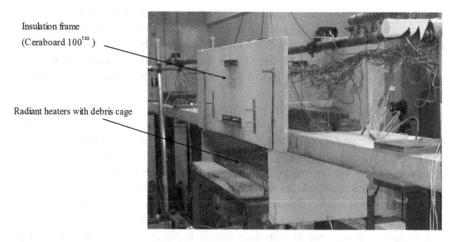

Fig. 4.9 Radiant heater assembly in use during test

4.4 Instrumentation

An instrumentation schematic is provided in Fig. 4.2. Four main types of measurements were of interest during the experiments: temperatures, tendon load (stress) levels, column strains, and slab deflections. A Vishay Measurements Group™ System 7000 data acquisition was used to record all sensor data from the instrumentation during testing.

4.4.1 Slab Temperatures

Slab temperatures are crucial for predicting unbonded prestressing steel tendon stress relaxation and the thermal/structural response of the slabs. Type K thermocouples were cast into each slab to provide thermal data. Thermocouples were placed at mid-span, with three thermocouples installed across the exposed soffit of the slab, two thermocouples within the concrete at the axis distance of the prestressing steel tendon, one thermocouple at the depth of the flexural reinforcing steel, and two thermocouples along the unexposed (top) surface. Unexposed surface temperatures of the slab were measured with thermocouples consisting of ceramic fiber wool and aluminum tape padding (see Fig. 4.10). Additional thermocouples were installed at the quarter points of the central span. One supporting column adjacent to the heated central span was monitored for temperature, with thermocouples placed along the head of the column. While care was taken to ensure instrumentation was functional for testing, some thermocouples could not be used due to the unavailability of sufficient data channels. The location of thermocouple instrumentation at mid span was verified after testing for all slabs and found to be within ±2 mm. All thermocouple locations are shown in Fig. 4.11.

A thermal imaging camera (FLIR A320A) was also used to record the spatial and temporal distribution of exposed soffit temperatures during heating. This system was used to gauge the uniformity of heating of the slab's exposed soffit, to identify pockets of high temperature, observe possible cracking, and to help indicate the time(s) at which any spalling of concrete cover occurred. The thermal imaging camera data were compared against the exposed soffit thermocouples. In a traditional standard fire furnace test it is currently difficult, or even impossible, to measure soffit temperatures in this manner. The furnace construction prohibits a full view of

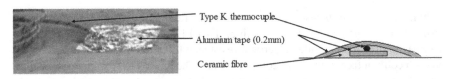

Fig. 4.10 Surface thermocouple treatment and placement

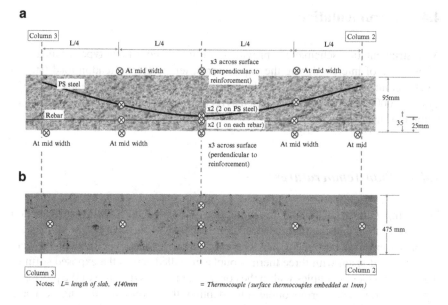

Fig. 4.11 Thermocouple locations shown for mid span of slab

the heated region and most furnace glasses prohibit this type of camera from measuring temperature. For this reason, the data from this camera will be invaluable for future computational modelling efforts and validation.

4.4.2   Tendon Stress Levels

Two through-hole load cells were installed at both the 'live' (jacked and anchored) and 'dead' (anchored) ends of the prestressing tendon in each slab to measure tendon stress levels before, during, and after fire testing. These were calibrated prior to, and after, the heating experiments. Tendon stress is indicated by dividing the average value of the load cells by the nominal tendon area, 93 mm^2.

4.4.3   Column Strains

Column strains were measured to give an indication of the resulting longitudinal restraint forces generated during testing. Two FLA-3-23 TML™ (3 mm gauge length) unidirectional bonded electrical resistance foil strain gauges were mounted on each column face to measure column strains. The gauges were mounted 750 mm from the base of the columns, half the distance between the top of the column cheek plates and the top plate (see Fig. 4.2). To verify the calibration of the gauges against temperature changes during the course of testing, temperatures were monitored adjacent to the strain gauge at Column 3 during every test. Temperature changes that could affect the strain gauges were compensated for using manufacturer specifications. The strain gauges installed on the supporting columns were calibrated by applying known loads to the columns with hydraulic jacks and measuring the resulting column strains at various locations. A load/unload exercise was performed on each column so recorded microstrains could be converted to restraining force (kN) with an R^2 value of more than 0.99. The column configuration was loaded and unloaded on the top plate. Restraining forces described herein are relative to that calibration. This allowed determination of the columns' true loading response.

Since the slabs were exposed to thermal gradients, there is no accurate way to determine the exact position of thrust against the supporting columns. For this reason axial forces in this chapter are estimated assuming thrust at the mid-depth of the column top plate. Strain gauges were placed within 2 mm of the location shown in Fig. 4.2.

4.4.4   Slab Deflections

Slab deflections were measured by five Celesco™ string-pot position transducers, which were supported by a protected instrumentation frame. The frame was installed above each slab before testing and acted as a support for cables from all sensors (see Fig. 4.2). The frame was protected from heating effects with one inch Ceraboard 100™ fiber boards (Fig. 4.9). One thermocouple was installed to monitor the temperature of the frame. The position transducers used had a measurement range of 127 mm. They were connected using heat resistant nickel-chromium wire attached to a bonded steel bolt at the slab surface, and placed at mid span and span quarter points of the central bay (see Fig. 4.2). The deflection at mid span of the outer bays was also monitored.

4.5   Procedure

This section describes the procedures used, from casting of the slabs, through testing, to slab decommissioning. Selected measurements are given when they affect the test procedures used.

4.5.1   Casting and Assembly

The as-constructed reinforcement diagram for Slab B is shown in Fig. 4.12. These images were taken prior to casting the concrete. A levelled steel channel bed was constructed below the formwork to guarantee that when the slabs were cast they would have the specified thickness along their full length. Slabs A and C were constructed identically to Slab B; however without tendon ducting or grout vents as indicated in Fig. 4.12.

To guarantee the as-designed tendon drape and concrete cover to the reinforcement, 12 diamond cut concrete chairs of variable size were installed below the reinforcement to support each non-prestressed bar. An additional total of 15 blocks were used to support the prestressing steel and plastic ducting along the length of the slab. Ten reinforcement links were added to support the negative moment tensile reinforcement. Steel wire (diameter 2 mm) was used to guarantee fixity of the steel and chair to the film-faced formwork (Fig. 4.13) and to prevent reinforcement from floating during concrete casting operations. Embedded thermocouples were fixed at the appropriate locations. Two vertical anchor tubes were cast at each column location. Lifting hooks were also installed at the slabs' quarter points. These were placed outside of the heated region.

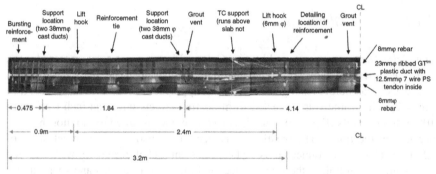

Notes: *Slab B is built symmetrical about the center line. Slabs A and C use one unbonded seven wire tendon rather than a cast plastic cast duct with tendon. All slabs are built identically within +/-5mm.*

Fig. 4.12 Slab B panoramic photo taken prior casting the concrete

Fig. 4.13 Unbonded tendon on a concrete chair tied down to form work

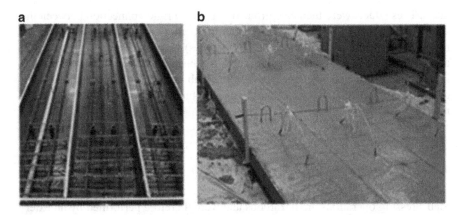

Fig. 4.14 (**a**) Slabs before casting and (**b**) slabs after casting

In order to monitor temperatures within Slab B's bonded prestressing duct during testing, a small hole was drilled in the top of the plastic duct near mid span, adjacent to the grouting vent, and two thermocouples were inserted. The precise positioning of these thermocouples was therefore unknown during testing; however, they were verified by post-testing excavation.

All three slabs were cast using ready mix RC40/50 concrete (see Fig. 4.14). The specified maximum aggregate size was 10 mm. Concrete cubes (100×100 mm) and cylinders (100 mm diameter $\times$ 200 mm tall) were cast concurrently and cured next to the test specimens. Three cubes and three cylinders were tested at 28 days in accordance to BS EN 12390 (BSI 2000) parts 1 to 8. The cubes had a compressive strength of 41 ± 2 MPa (average $\pm$ standard deviation), whereas the cylinders had a split cylinder tensile strength of 2.8 ± 0.5 MPa. The non-prestressed 8 mm diameter reinforcement tested to an ultimate strength of 658 ± 7 MPa, and the prestressing tendon's center wire was tested to 2033 ± 11 MPa ultimate strength. The slabs were cast in November 2011 and tested in June 2012, allowing over 6 months curing time. Cube strength tests were performed when the slabs were tested with strength averaging close to 50 MPa.

4.5.2 Lifting

The slabs were lifted onto their supports according to lifting guidelines set out by the CPCI (2007) design manual, using four hoist points made out of 6 mm diameter mild steel reinforcement. A layer of approximately 10 mm deep cementitious grout, consisting of one part Portland cement, three parts siliceous sand by mass, and a water-to-cement ratio of 0.4, was placed on each column head prior to seating. The slab was then placed on this grout and enabled to settle naturally. For each column-slab connection, two 20 M Grade 10.9 bolts were inserted into the precast column ducts. These ducts were then also grouted to create a rigid connection between the slab, column, and bolt. A top layer of grout of about 1 cm thickness was added, a rigid steel plate (identical to the column top plates) was placed on top of the grout, and the bolts were tightened. This process was repeated for all four columns, creating four semi-rigid connections between the slab and the column supports. This created conditions that would as closely as possible simulate those found in a real multi-bay building.

4.5.3   Tendon Stressing

Slab stressing operations are hazardous and were therefore performed by an external contractor, ConForce UK, who did so as an in-kind contribution to the project. The tendons in all slabs were stressed to approximately 1341±7 MPa (about 125 kN). Over 1 h, the slab was stressed in three ramps of approximately 450 MPa each. When the target load (near 70 % of the tendons' ultimate tensile strength) was achieved, the prestressing anchors were set and the jacks released and removed. The subsequent phase of testing did not commence until prestress loss differences during any consecutive 24 h interval was less than 0.5 MPa. This was necessary so that short-term creep losses in the slabs would not influence the heating tests.

The unbonded slabs were then loaded, whereas the bonded slab was grouted prior to loading. The bonded PT slab was grouted with Parex™ cable grout (65 MPa compressive strength after 28 days) inside the embedded plastic monostrand cable duct. This work was also performed with the assistance of ConForce UK. Loading commenced 2 weeks after grouting in this case. For all slabs, after seating and short term losses, average tendon stress stabilized near 1165±13 MPa on the day of testing—this is a realistic in-service stress level for post-tensioned tendons in a real building.

4.5.4   Loading

Slabs were loaded with lead brick weights. These bricks were used because they can remain stationary during small deflections and be directly exposed to temperatures below 300 °C. Each lead brick was weighed, sequentially numbered, and placed identically for each slab. Placement of weights was as shown in Fig. 4.1. The expected theoretical deflection from the lead weights at mid span for the continuous slab was predicted to be 2.1 mm. The actual short term deflection prior to testing was measured as 1.6±0.5 mm. Less than 10 MPa of tendon stress increase (due to tendon elongation from deflection) was observed in any test during or after loading. In all cases there was a minimum 1 week interval between loading and heating.

4.5.5 Heating

Slabs A and B were heated locally with the radiant heaters until the presumed 'critical temperature' of the prestressing steel tendon was reached, according to EN 1992-1-2 (CEN 2004) (i.e. until 350 °C was recorded at the tendon location). This resulted in total heating times of approximately 200 and 215 min for slabs A and B, respectively. Slab C was heated until a temperature of 427 °C was measured at the tendon, in accordance with the higher critical temperature used in North American guidance (IBC 2012). This resulted in a total heating time of 270 min for Slab C. After these criteria were achieved the propane supply was turned off and the radiant heaters were removed from beneath the slabs to allow them to cool to ambient temperature. Each slab was continuously monitored during both the heating and cooling phases, for a total of 24 h.

4.5.6 Post-test Examination

The true ultimate capacities of the PT slabs at ambient condition were unknown making a residual capacity test to failure of little scientific value for comparison purposes. Therefore the slabs were not loaded to failure after heating. The unloading behavior of the slabs and residual deflections after the applied load was removed indicate the presence of plastic deformation. After unloading, slabs A and C were de-tensioned. All parameters were monitored during de-stressing and the amount of post-test camber was compared to the original camber imposed during the prestressing operations. The tendons were then extracted and analyzed for their residual properties. Finally the slabs were cut in two and removed from the laboratory. Immediately when the slabs were taken outside, the thermocouple placements were then verified by excavating them from the concrete. Four moisture content samples were extracted in the anchorage zones of both ends of the slab (the unheated portions of the slab). The grout from Slab B was also tested for moisture content. All moisture contents were determined by dehydration mass loss following heating at 120 °C for 2 days.

4.5.7 Safety Considerations

Post-tensioning operations have inherent risks. A number of measures were taken during the experiments to ensure the safety of all personnel involved. The primary safety concern was the potential for tensile rupture of the stressed prestressing steel while testing. An additional concern was the safety of individuals near the radiant heater system during testing. As such, and despite the best efforts to ensure proper instrumentation performance throughout the tests, no risks were taken to re-apply or modify instrumentation near the live tendons or radiant heaters during testing.

This was an issue during testing of Slab C for which two strain gauges malfunctioned on Column 1 near the anchorage at the start of the test. Since no attempt was made to remedy this during the test, the data had been lost. Additionally, because of the risk of slab collapse during cooling, the radiant heaters were removed from underneath the slab during cooling, thereby blocking the thermal imaging camera from measuring this interesting temperature response.

4.6 Observations

All three tests showed a highly complex, similar, and in several respects unexpected, deflection response during high temperature exposure, despite the structural system's relatively simple construction. For all stages of high temperature testing, a summary of the measured deflections for Slab A, B and C is noted numerically in Fig. 4.15 (with the deflection profile artificially amplified for clarity and shown relative to deflections at the onset of heating). The figure shows several phases of slab deformation that occurred during testing (phases denoted 1–3 in heating and 4–5 in cooling). All slabs exhibited a similar response to heating and cooling. Figure 4.15 shows that deflection values were generally small (peaking at approximately 10 mm or span/400).

Testing also revealed that when heated locally, the slabs did not follow a typical gravity deflection profile for the central span. Rather, deflection was primarily influenced by thermal bowing brought on from differential thermal gradients in the through-thickness.

Figures 4.16, 4.17, and 4.18 summarize the slabs' thermal and structural responses with time during both heating and cooling (measurements taken at mid-span). For rapid comparison, these figures present the mid-span vertical deflection relative to start time of heating, maximum thermal gradient (the difference in temperature between the heated face and the cool face), exposed soffit surface temperature, and prestressing tendon (axis distance) temperature with time from the onset of heating. The average lateral restraining force developed on the four columns, measured by instrumenting and calibrating the vertical cantilever column supports with electrical resistance strain gauges, as well as the prestressing tendon stress level measured at the end anchorages, are shown for all tests.

While such a complex and interesting deflection response has not been previously observed in traditional 'pass-fail' testing of simply supported elements tested in furnaces, they have considerable importance for demonstrating a credible ability to model real PT concrete structures when exposed to fire. A discussion of, and possible explanations for, the observed physical mechanisms that might be responsible are therefore given in this section. These measurements are considered in greater detail within the next section.

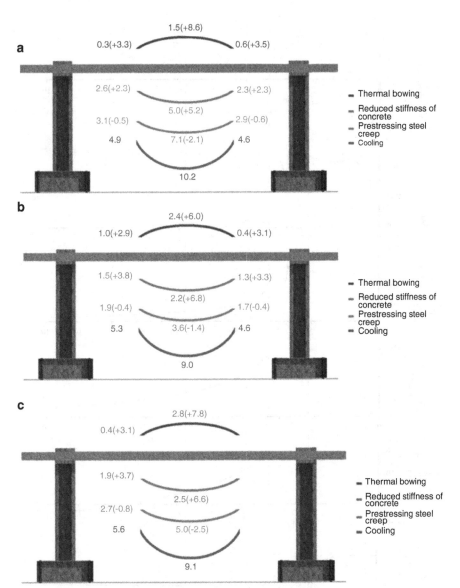

Fig. 4.15 Deflection profiles for center span as measured by position transducers for first heating of PT slabs: (**a**) Slab A, (**b**) Slab B, and (**c**) Slab C. *Notes:* + denotes camber, – denotes deflection. Deflection profiles are interpolated between gauges. Profiles not to scale. One position transducer in Slab C failed during testing, therefore these values are not shown (values in mm)

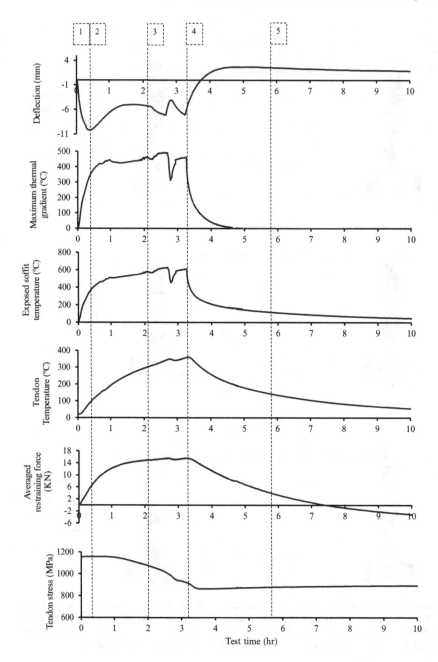

Fig. 4.16 Slab A behavioral response (unbonded)

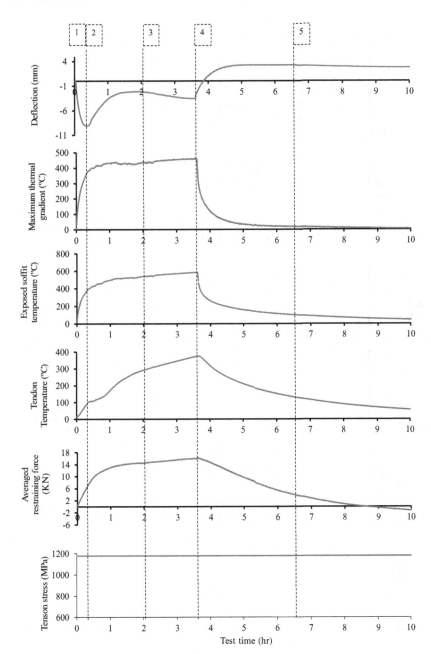

Fig. 4.17 Slab B behavioral response (bonded)

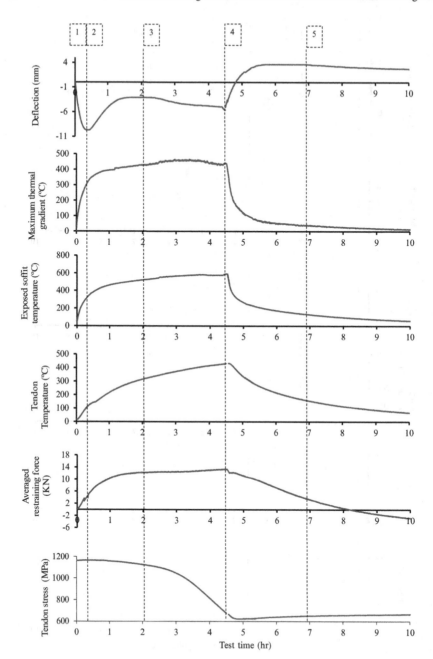

Fig. 4.18 Slab C behavioral response (unbonded)

4.7 Interpreting the Structural and Thermal Response of the Tests

4.7.1 Exposed Soffit Temperature Observations

The repeatability of heating exposure for the tests is crucial for making comparisons between the respective slab responses. For this reason, the consistent performance of the radiant panel assembly (both spatial and temporal) between tests was important. A comparison of recorded exposed soffit temperatures in slabs A, B and C is useful to compare the consistency of the heating provided.

Table 4.2 presents 3 h of temperature-time data for the slabs' A, B and C soffit at mid span. Averages within the table are calculated from the exposed soffit thermocouples for each test. This represents three thermocouples for Slab A and C, and two thermocouples for Slab B, as one its thermocouple failed during testing. A full average of all eight thermocouples and the resulting standard deviation is also tabulated. In general there was reasonable consistency of heating between all three tests, confirming the repeatability of testing using the radiant heater assembly.

To further explore possible spatial distributions of temperatures along the exposed soffit, a thermal imaging camera was used. The thermal imaging data were also compared to the thermocouple data for verification purposes. It was expected that there might be some non-uniformity in heating; however, the maximum standard deviation observed along the length of the prestressing steel tendon according to the thermal imaging data was never greater than 40 °C at any time (see Table 4.3).

Table 4.2 Comparison of exposed soffit temperatures at mid span

Time (min)	Slab A (°C) Average	Deviation	Slab B (°C)[a] Average	Deviation	Slab C (°C) Average	Deviation	Average results (°C)[b]
30	426	10	428	43	379	7	409±21
60	498	29	496	46	454	7	481±32
90	522	24	525	29	488	7	510±25
120	559	47	539	27	511	9	536±36
150	602	45	557	21	532	16	565±42
180	593[c]	–	571	17	564	48	–

[a]Data for Slab B is based on two thermocouples due to an instrument malfunction
[b]Average and standard deviation data is based on eight thermocouples for Tests A–C
[c]Data is based on only one thermocouple so no deviation could be calculated

Table 4.3 Averaged exposed soffit temperatures using thermal imaging camera at mid span

Time (min)	Slab A (°C) Average	Deviation	Slab B (°C) Average	Deviation	Slab C (°C) Average	Deviation
30	442	31	446	31	435	32
60	522	34	502	39	487	27
120	567	40	530	24	526	28
180	585	37	551	23	559	32

Notes: Values are representative of nine equally spread points along the heated portion of the soffit

4.7.2 Unexposed Surface Temperatures

Unexposed surface temperatures can be used to highlight the thermal gradient through the slab, which would promote thermal bowing during heating.

With reference to the preceding discussion of exposed soffit temperatures, an increasing temperature difference can be observed as the tests progress. For all tests this difference eventually exceeds 450 °C, thus considerable thermal bowing should be expected. Figures 4.16, 4.17, and 4.18 illustrates this maximum thermal gradient in the slabs as monitored from the start of the test to the end of cooling. This measurement is constructed on the basis of the difference between the middle most thermocouple readings taken on the unexposed surface and the exposed surface at mid span.

4.7.3 Reinforcement Temperatures

Figures 4.16, 4.17, and 4.18 show the average of the two prestressing steel tendon temperatures, taken from thermocouples on the tendon at mid span, for all three slabs. It is assumed that at mid span the highest tendon temperature was recorded, since this is where maximum tendon drape occurred and was the most heated location in terms of both radiative view factor and direct convective heating from the radiant panels. The deviation between the two tendon thermocouples in each slab never exceeded ±2.2 °C (essentially the accuracy of a Type K thermocouple).

Slab B demonstrated a reduction in heating rate in the temperature region near 100 °C, which is thought to be due to the measured higher moisture content of the cementitious grout used to create a bonded PT situation within the plastic post-tensioning duct. The moisture content of the grout was 6 % by mass, as compared with the surrounding concrete which was about 4 %.

Maximum temperatures during the heating for Slabs A and B reached more than 350 °C, the critical temperature given by EN 1992-1-2 (CEN 2004). In Slab C, the prestressing tendons were heated to more than 427 °C, the critical temperature given by IBC (2012).

Reinforcement temperatures were slightly higher than those recorded at the tendons, although these temperatures never exceeded 500 °C during any test. The slab design called for these bars to be 400 MPa in ultimate strength at ambient. The reinforcement was tested in ambient at 658±7 MPa using reduction factors at 500 °C (CEN, 2004), assuring that there would have been more than 400 MPa in tensile strength remaining at high temperature.

4.7.4 Tendon Stress Levels

The relationship between tendon stress and temperature is critical for UPT structures. Under certain combinations of stress and temperature tendons can experience tensile rupture due to heating alone. The variation of tendon stress with the tendon temperature measured at mid span is plotted in Figs. 4.16, 4.17, and 4.18

for Slabs A and C. Since Slab B had a grouted and bonded tendon, no measureable stress relaxation was observed during heating by the load cells installed at the tendon anchorages.

Figures 4.16, 4.17, and 4.18 show considerable stress relaxation for both slabs with unbonded tendons. Obviously this is due to a combination of thermal and structural actions, including thermal elongation of the tendon, thermal expansion of the concrete, thermal bowing of the slab, reductions in elastic modulus of the tendon at elevated temperature, and creep of the tendon. Reductions in tendon stress accelerate with increasing temperature. Even once the heating stopped, these reductions continue due to the thermal inertia in the concrete. A minimal amount of stress recovery is evident during the cooling phase, indicating that considerable irrecoverable creep of the tendons had occurred. There is potential for irrecoverable creep in unbonded tendons within concrete in fire (postulated in Gales et al. 2011a, b), this is confirmed by the data in Figs. 4.19 and 4.20.

The observed stress relaxation, or lack of stress relaxation, would have an impact on a structure's ability to balance the deflection and create other inter-related consequences on the PT slabs. Relative restraining forces and deflection of the slab would be directly impacted by stress relaxation. This is because the prestressing steel tendon's pre-compressive action on the slab, and its ability to balance applied loading, would diminish upon heating. Whether such structural actions could be defensibly modelled using the best available finite element structural modeling codes, even for this simple case with a very well defined thermal profile, remains an open question.

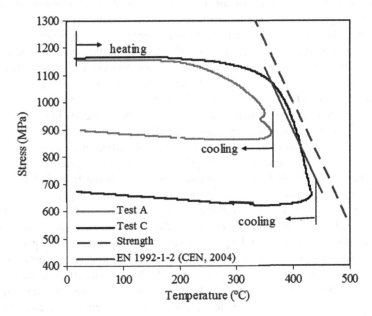

Fig. 4.19 Unbonded prestressing steel stress relaxation with respect to measured tendon temperature

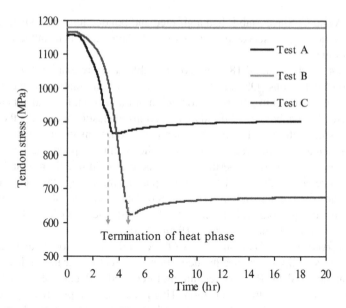

Fig. 4.20 Prestressing steel stress relaxation with respect to time for all slabs

As already stated, one issue of concern in UPT slabs during fire is sudden tensile rupture of the prestressing steel tendon due to localized heating. There is data on the reduction in strength of cold-drawn prestressing steel wire with increasing temperature in the literature, for example in EN 1992-1-2 (CEN 2004) which is plotted in Fig. 4.19. In addition, separate strength tests were performed to characterize the prestressing steel used in these tests. In Fig. 4.19 it can be seen that Slab A was never at serious risk for tensile failure of the tendon during testing.

Slab C included the aforementioned disk-spring anchorage to simulate a longer tendon length and thus a shorter heated length ratio (~10 % instead of 17 %). Plotting the stress relaxation-temperature results for Slab C in comparison to Slab A confirms that for shorter heated length ratios less recoverable relaxation will occur. The data for Slab C indicate an interception with the EN 1992-1-2 (CEN 2004) strength reduction curve for prestressing steel at high temperature, however, the tendon stress does not intercept the strength reductions extrapolated from the data produced for Fig. 4.19 (Gales and Green 2015). Post fire evaluation of the slabs confirmed discrete necking on the prestressing steel tendon, representing 10 % area reduction and indicating that the tendon was indeed on the verge of failure during this test. This gives some credibility to the EN 1992-1-2 (CEN 2004) reduction factors as a conservative means to predict tendon failure under stress at high temperature. It must be noted that heated length ratios below 10 % represent the more likely fire scenario in a real building, and these were not tested. This would be similar to the behavior of a multi bay structure of 40 m in total length heated in a single bay, or compartment. For larger structural configurations with total floor plate dimensions greater than 40 m, smaller heated length ratios should be expected in practice. In the case of Slab C, had the spring simulated an even greater length of unbonded tendon to mimic just one additional structural bay, the tendon likely would have ruptured.

The above discussion has not considered the possible effects of more rapid heating, which would result in less time-dependent prestress losses and possibly an earlier stress-strength interception resulting in tendon rupture.

4.7.5 Restraining Forces

It is well known that restraining forces and secondary load carrying mechanisms may benefit the performance of structural systems during fire. However, there are cases where high restraining forces can increase the formation of perimeter cracking, column shear loading and even failure, and/or slab spalling due to induced compressive stresses. Restraining forces are generated based on the boundary conditions at the supports and the stiffness of the surrounding structure.

Each supporting column had two identically placed calibrated strain gauges allowing them to be used effectively as load cells (see Fig. 4.2). A number of assumptions were made to estimate the applied force on the column head from measured strain. First, it was assumed that the columns were at uniform temperature, given that they were not directly exposed to heating. Second, it was assumed that the total strain measured on the column was the sum of the flexural and axial strain components. Through the aforementioned calibration exercise, the flexural strains were converted into an applied reaction force. Since fluctuations in temperature can exist within the laboratory during testing (for every test these were monitored on Column 3), the axial strain may be influenced by the temperature of the column. Therefore the normal reaction force applied vertically on the columns could not be accurately measured. The horizontal thrust/restraint measurement is assumed not to be influenced by temperature changes in the laboratory by assuming that temperature change is uniform in the supporting columns.

Table 4.4 gives the average thrust exerted by each slab on all columns (1–4) during heating and thermal pull (contraction) after cooling, for all tests. The averages for Slab C are compiled only on the basis of three columns (2–4), since Column 1's

Table 4.4 Maximum thrusts and pulls on columns

Test #	Maximum thrust in heating (kN)	Relative compressive stress difference for maximum thrust (MPa)	Maximum pull in cooling (kN)
A	15.7 ± 3.0	0.34 ± 0.07	4.8 ± 1.7
B	16.3 ± 2.8	0.36 ± 0.06	3.4 ± 1.6
C	13.4 ± 2.8	0.29 ± 0.06	5.2 ± 2.0

strain gauges failed during testing. The thrust is resolved into stress by considering the full cross-sectional area of the slab. These stresses were less than 0.5 MPa in all cases. A consistent response in most tests indicated that all columns experienced a similar thrust-contraction response.

It is important to note that the observed restraint forces apply for one boundary (heat and support) condition only. These forces were relatively small during the current tests, but they are only representative of a thin strip of slab as tested. When considered for a real two-way slab of the same thickness placed on an arbitrary 4×4 m grid, heated within a center bay over half its area, a restraining force of 100 kN or more could easily be exerted on each column, assuming a exerted stress of 0.5 MPa.

Had more of the slabs' soffit been heated, or had the slabs been heated to a higher temperature, it is possible that larger thermal thrusts would have been induced. This raises important questions about the shear capacity of supporting concrete columns and their ability to resist lateral and axial forces to prevent progressive collapse mechanisms. For rational design in a performance based environment, the likely fire scenario and likely end/intermediate support conditions must therefore be known.

4.7.6   Deflections

To date no experiments have carefully described the deflection behavior of a restrained and continuous one-way PT slab during localized heating. The deflection response is crucial to identify and understand various interacting mechanisms occurring in the structural system during fire, and also for validating computational models of PT slabs during fire.

Figures 4.16, 4.17, and 4.18 show the respective vertical deflections at the mid points of slabs A, B and C during testing, with deflections zeroed at the onset of heating in relative comparison to other observed behavior. In all cases quarter point deflections were approximately one half the mid span values. The unheated end spans displayed negligible deflections. All tests displayed a five phased distinct deflection trends during heating.

Phase 1: Thermal Bowing The similarities in response of each slab despite the differences in prestress characteristics were striking. In all tests, rapid downward deflection was initially observed: nearly 10 mm in the first 30 min of heating. Figures 4.16, 4.17, and 4.18 show that this phase of deflection is likely dependent on thermal bowing of the concrete. This is influenced by increasing thermal gradients with time (>350 °C) and the resulting differential thermal expansion of the concrete through its thickness. The resulting restraining forces measured during this phase also increased with the development of the thermal gradient. During this phase, prestressing tendon temperatures were observed to peak around 100 °C in all slabs, leading to the expectation of little stress relaxation. Indeed, a small increase in tendon stress due to mechanical re-stressing of the prestressing tendons in Slabs

A and C from bowing and slab elongation would be expected, and approximately 10 MPa in both unbonded slabs A and C was observed. This phase ends when deflection reverses and the slabs begin to deflect upwards.

Phase 2: Concrete Mechanical Deterioration and Load Induced Thermal Strain (LITS) The second phase is the upward deflection, or cambering, of 5–7 mm in all slabs. It is less easily explained than Phase 1. Figures 4.16, 4.17, and 4.18 indicate that cambering is observed for the duration of this phase despite an increasing thermal gradient for the full duration of this phase. The gradient reaches 420–435 °C (divided by 95 mm the slab thickness) by the end of the phase and would lead to an expectation of additional thermal bowing and downward deflection in the absence of any prestressing force. An increase in restraining force also continues to closely follow the trends of an observed temperature gradient increase; however, the slabs do not continue to deflect downward. It is clear that multiple interrelated thermal and physical mechanisms must be occurring to cause this response. This cambering response may be influenced by loss of stiffness of the heated concrete near the heated face. The exposed surface of the concrete in all cases in this phase ranges from 400 to 550 °C. Concrete is known to suffer substantial stiffness losses at these temperatures. This stiffness loss supports an upward movement of the effective neutral axis of bending due to the eccentric prestress changes within the heated region. With the loss of stiffness and the maintenance of high prestressing levels in the tendon it would be expected that the slabs would camber during this phase in testing. The measured prestress relaxation was at most 10 % by the end of this phase for slabs A and C. Under the high initial compressive forces induced on the slab from post-tensioning, the upward movement of the slab can also be influenced by the occurrence LITS of the concrete within the heated region (see Khoury 2006, for further discussion of LITS). While minor in this case, the deflection response is further complicated by a possible shift in tendon eccentricity due to the melting of the polypropylene extrusion sheath in slabs A and C, or the plastic corrugated duct in Slab B. This could occur when tendon axis distance temperatures exceed 100 °C, and would decrease the tendon eccentricity by at most 2 mm. This eccentricity change would have had a very small contribution to increasing downward deflection, rather than cause camber. This phase is considered to have ended when slab cambering transitioned once again into a downward deflection trend.

Phase 3: Prestressing Steel Tendon Creep Causing Relaxation Cambering ceased in the previous phase, followed once again by downward deflections when the prestressing tendon temperatures neared 300 °C. This deflection was small, approximately 2 mm for all slabs. In this phase, the deflection trend resembles the beginning of a traditional creep curve for a steel prestressing tendon. This appears as the result tendon stress relaxation at higher temperatures due to thermal creep elongation. This stress relaxation causes an effective loss of pre-compression of the slab. Gales et al. (2012) previously identified that most modern prestressing tendons show rapidly accelerating creep at temperatures above 300 °C. The resulting reductions in tendon stress can cause global increases in deflection for the PT slabs. This deflection cancels out the cambering effects of the locally heated region that occurred in Phase 2. During Phase 3, the thermal gradient is still observed; however, it is almost stationary and the rate of this increase is diminished significantly. Subsequently, additional deflection anticipated from thermal bowing is considered to be small. In Slab B there is a possibility that the bond between the tendon and grout had deteriorated, effectively making a section of the PT reinforcement unbonded. This effect would create localized tendon stress relaxation zones within the slab. A post-test evaluation of Slab B identified a region of slough off cover spalling where significant cracking in the post tensioning grout was confirmed. This indicates that the bond may be compromised and that localized relaxation of the tendon was permitted. In the test on Slab A the radiant heaters failed briefly due to 'blowback', this can be seen in Fig. 4.16 with the sudden deflection response of the slab in cooling, immediately and profoundly effecting nearly all measurements. This clearly demonstrates the importance of differential thermal gradients on concrete structures. It is well known that in steel structures thermal deflections are far more significant than mechanical 'load induced' deflections; however, these same considerations seem to go unacknowledged in concrete structures. When the heaters failed, cooling of the slab began causing cambering of the slab due to a decrease of the thermal gradient. When heaters were re-ignited and the slab was again under increasing thermal gradient, it began to deflect and resumed its previous deflection response. None of the slabs were taken to 'failure' (i.e. structural collapse). Heating was terminated once the prestressing steel tendons reached temperatures considered to be 'critical' (350 °C as in Europe for slabs A and B, and 427 °C as in USA for Slab C).

Phase 4: Cooling, Recovery Upon halting the heat exposure, the cooling phase was monitored for several hours (during phases 4 and 5). During these stages all slabs reversed their deflections, cambering approximately 7–9 mm after cooling. This phase appears to be controlled by thermal contraction due to reductions of thermal gradient. This behavior was briefly observed for Slab A in Phase 3 during accidental cooling as noted above. At the beginning of Phase 4, for all slabs, the prestressing steel tendon continued to relax in prestress, despite the overall cooling of the slab. This may be influenced by accelerating creep damage of the prestressing tendon as its temperatures still exceeded 300 °C, as well as the 'thermal wave' which will continue in the concrete even after the heating is removed. Once the prestressing tendon temperature dropped below 300 °C this prestress loss stopped and subsequently the tendon began to regain prestress as a result of thermal contraction of the cooling steel. This would have also influenced the slabs' deflections. Rather than arch the slab up in camber due to tendon stress increase, the slab once again begins to show a new deflection phase. Phase 5 begins to occur when the exposed soffit of the slabs cooled to near 100 °C.

Phase 5: Cooling, Rehydration The slabs all began to deflect down slightly at this stage of testing. This deflection behavior could be explained by the slabs absorbing moisture from the atmosphere and rehydrating the lime in the Portland cement. This action, however minor, could cause an expansive effect on the slab and act to deflect the slab downward.

Interestingly, all slabs exhibited a relative camber after cooling. At this stage, the residual deflection of all slabs was up relative their pre-heating condition. All slabs indicated a gradual decrease in the relative restraining force exerted on the columns, settling on a relative residual 'pull-in' restraining force on all supporting columns. They all also had less recovery of deflection when compared to the thermal bowing observed in Phase 1. These behaviors indicate permanent plastic deformation of the entire structural system and could potentially have significant consequences on a post-fire assessment of structural stability in real buildings. This means the assessment of UPT buildings cannot necessarily rely on vertical slab deflection to provide an indication of structural damage due to heating. Slabs A and C exhibited more deflection compared to Slab B, this likely due to prestress relaxation in the unbonded tendons for these two slabs. Since Slab B was bonded, it presumably maintained a greater proportion of its initial prestress and thus experienced the least overall deflection throughout testing.

For every fire and every PT structural configuration that might exist in a real UPT building, these five phase deflection response clearly cannot be assumed for all cases. As suggested above, the deformation response of the slab is controlled by a number of test variables as well as thermal and physical mechanisms. For instance, the severity and uniformity of heating induced on the slabs will play a highly significant role on the degradation of concrete properties and the in-service stress level of the post-tensioning steel. Different structural dimensions, PT tendon drape, loading level, restraining frame stiffness, and two-way reinforcement and prestressing (membrane actions) would all influence the observed response in a real building in ways not yet understood. A different deflection response should be expected if experimental procedures are even slightly varied. In the absence of additional testing, modelling may shed light on the response of these structures in different structural configurations, assuming that modelling can demonstrate an ability to account for the complexities observed in these tests.

4.8 Post-heating Evaluation

A post-heating evaluation was conducted for each slab. Each slab was analyzed for cracking and spalling patterns. Figures 4.21, 4.22, and 4.23 illustrate these observations. This discussion focuses primarily on Slab C. Discussion of slabs A and B is limited because these slabs were heated a second time to investigate LITS and thus some the post-heating evaluation conclusions are not applicable. The post-heating analysis of Slab C focusses on the prestressing steel and its deflection behavior at various stages of decommissioning (unloading, distressing, etc.). For all slabs, the final evaluation included excavation to determine thermocouple placement, and concrete cover depth at critical locations, and confirm a satisfactory tendon drape. Selected observations in relation to the slabs are discussed below.

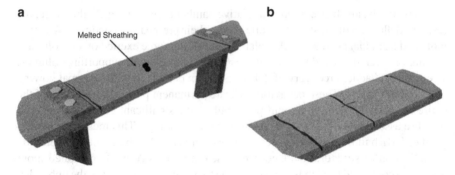

Fig. 4.21 Cracking and heating pattern of Slab A center span (**a**) unexposed surface view, and (**b**) exposed surface view (upturned)

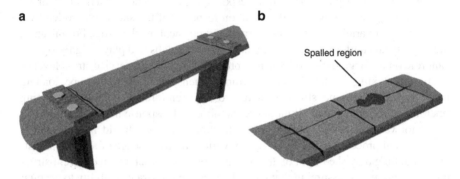

Fig. 4.22 Cracking and heating pattern of Slab B center span (**a**) unexposed surface view, and (**b**) exposed surface view (upturned)

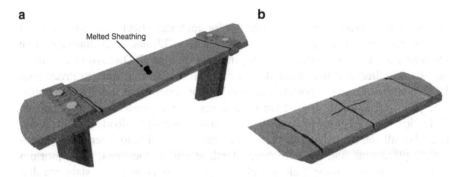

Fig. 4.23 Cracking and heating pattern of Slab C center span (**a**) unexposed surface view, (**b**) exposed surface view (upturned)

Fig. 4.24 Discrete
transverse cracking at
column location (Slab A)

4.8.1 Slab A

Spalling was not present on the exposed soffit of Slab A. During testing, black smoke was observed at the top of the unexposed surface. This was thought to be from the prestressing steel sheathing burning, and the smoke emerging through the embedded thermocouples wire channels. Small transverse cracks were also seen along the exposed soffit of the slab within the heated region.

Transverse cracking, with a widths of less than 1 mm and depths extending to the upper layer of mild steel reinforcement, was present at each support (adjacent to Columns 2 and 3). These cracks were present in all slabs (see Fig. 4.24).

After cutting the slab equally in two, it was sectioned using a diamond saw. The concrete cover was confirmed within 2 mm of the design at all locations. A layer of approximately 10 mm of pink concrete was identified at the exposed surface of the slab's soffit. Pink concrete is accepted opinion to occur at temperatures greater than 300 °C (Hager 2014) and is a key indicator of strength loss. Color changes in concrete are dependent on exposure temperature, occurring due to the oxidization of iron in the cement (Ingham 2009), so it can indicate that the concrete has been damaged by phase changes in the cement. Slabs B and C also exhibited the same 'pink' zone depth on examination.

4.8.2   Slab B

Very mild spalling, also known as sloughing, was noted during testing, both visually and through the thermal imaging camera. Figure 4.25 illustrates when spalling of the concrete first occurred, at 8 min into testing. This location below, shown in Fig. 4.25, the tendon was examined post-test, and a small (<5 mm wide, by 5 mm deep) section of concrete was missing. A second spalling location within the heated region was observed shown in Fig. 4.26. Post-heating examination indicated that the

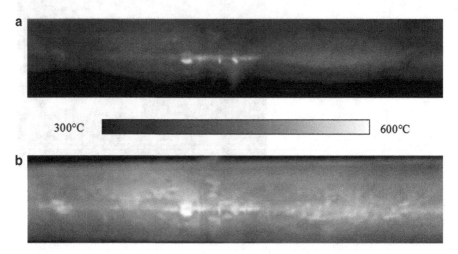

Fig. 4.25 Two thermal imaging camera images taken during testing in Slab B (**a**) at 8 min, and (**b**) at 2 h

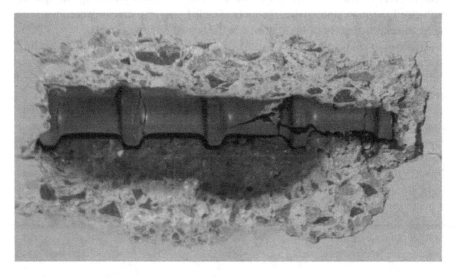

Fig. 4.26 Exposed section of grouted duct after Slab B on exposed soffit

prestressing steel grout duct was exposed at the spalling location and that the poly-
mer duct had melted away. Discrete cracks were observed near the location of the
duct's corrugated ribs. These may have been caused by the removal of the duct by
melting and then the tendon's restrained movement in heating. Figure 4.26 illustrates these cracks.

Transverse and longitudinal cracking (<1 mm) were both observed. Longitudinal
cracks are formed due to the longitudinal expansion of the tendon, which relives
compressive forces on the concrete, then the tensile forces on concrete can split
weaker sections near tendons. The crack appeared on both the exposed soffit and the
top surface of the slab. It was only in the heated region, extending approximately
10 cm in the cold zone on both ends, and was coincident with the location of the
prestressing tendon. Such longitudinal cracks have been observed in many previous
tests of post-tensioned concrete elements during fire. This supports the hypothesis
that under a local fires, longitudinal cracking should arrest where the prestressing
steel is no longer exposed to high temperature. This would be due to the relaxation
of prestress in one location, and maintenance of the precompression outside the
heated location. Longitudinal cracking was only visible in Slab B. This was likely
due to the larger duct size (23 mm) and may have been caused by the smaller effective concrete section's ability to resist tensile stresses. Full grouting of the post-
tensioning duct was confirmed within the center and adjacent spans.

4.8.3 Slab C

Discrete transverse cracking was present on the slab at mid span and at the supports
as shown in Fig. 4.24. Some discrete longitudinal cracks were confirmed along the
exposed soffit of Slab C below the location of the tendon; however, they were very
fine and barely noticeable.

As with Slab A, no spalling or cover separation was identified on Slab C. When
unloading Slab C, a camber of 2.5 mm was measured at mid span and a subsequent
deflection of 3.2 mm was observed when de-stressing the tendon. These values were
greater than the original camber (1.8 mm) and deflection (1.2 mm). These are
thought to be a result of the degradation of material properties and irreversible
strains in the concrete during heating.

During testing, flaming was observed on the slab soffit parallel to the tendon.
This was caused by the tendon grease inside the prestressing sheath igniting as it
seeped through the surface of the slab.

Once de-stressed, the prestressing tendon was extracted. A portion of the tendon
was used to assess the diamond pyramid hardness to indicate tensile strength. The
diamond pyramid hardness was measured at 462, indicating strength of 1666 MPa
and verifying a peak temperature in the range of 400–450 °C, in line with the
strength-hardness correlations given by Roberston et al. (2013). The remaining
portion of the tendon was tested and gave an ultimate strength of 1685 MPa.

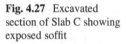

Fig. 4.27 Excavated section of Slab C showing exposed soffit

An excavated section of Slab C is shown in Fig. 4.27, it shows residual sheathing traces through the depth of the concrete. Also shown in Fig. 4.27 is the presence of pink concrete.

4.9 Summary

The main objective of the experiments described was to investigate the thermal and structural response of multiple span continuous, restrained one-way PT concrete slabs exposed to localized heating. Despite the comparatively simple layout of these tests and the well-defined uniform heating applied to the slabs' exposed soffits, the resulting global response is extremely complex. The tests demonstrate numerous thermal and mechanical actions which would need to be carefully considered to rationally model even these extremely simple PT concrete structures. These tests will assist future efforts to build rational and defensible models that can account for the requisite complexities. In the meantime, several conclusions can be formulated on the basis of the test results:

- Considerable prestress losses occurred during heating. The magnitude of stress relaxation was primarily influenced by the heated length ratio, as well as time, temperature, and loading conditions;
- Premature tendon rupture is a realistic and likely scenario for smaller heated length ratios in PT concrete slabs;

- Restraining forces will be generated when a PT slab is exposed to localized heating, these may be of sufficient magnitude to cause distress to supporting columns; and
- The continuous and restrained, one-way spanning PT slabs tested exhibited four distinct phases of deflection, as described previously. These trends are influenced by a complex interplay between stiffness degradation of concrete and prestressing steel tendons, plastic and creep deformation, and thermal expansion/contraction. Any credible attempt to numerically model such systems in fire must account for this complexity.

References

Allouche, E. N., Campbell, T. I., Green, M. F., & Soudki, K. A. (1998). Tendon stress in continuous unbonded prestressed concrete members – Part 1: Review of literature. *PCI Journal, 43*(6), 86–93.

Allouche, E. N., Campbell, T. I., Green, M. F., & Soudki, K. A. (1999). Tendon stress in continuous unbonded prestressed concrete members – Part 2: Parametric study. *PCI Journal, 44*(1), 60–73.

ASTM. (2014). *Test method E119-01: standard methods of fire test of building construction and materials* (Rep. No. E119-14) West Conshohocken, PA: American Society for Testing and Materials

Bletzacker, R. W. (1967). Fire resistance of protected steel beam floor and roof assemblies as affected by structural restraint. *Symposium on Fire test methods. American Society of Testing Materials*, pp. 63–90.

BSi. (1997). *Structural use of concrete. BS 8110-1*. London: British Standards Institution.

BSI. (2000). *Testing hardened concrete. BS 12390-1-8*. London: British Standards Institution.

Buchanan, A. H. (2001). *Structural design for fire safety*. New York, NY: Wiley.

Canadian Standards Association. (2004). *CAN/CSA A23.3-04: Design of concrete structures*. Ottawa, ON: CSA.

CEN. (2004). *Eurocode 2: Design of concrete structures, Parts 1–2: General rules-structural fire design, EN 1992-1-2*. Brussels: European Committee for Standardization.

CEN. (2005). *Eurocode 3: Steel design, Parts 1–8: Joints and welds, EN 1993-1-8*. Brussels: European Committee for Standardization.

CPCI. (2007). *Design manual: Precast and prestressed concrete*. Ottawa, ON: Canadian Prestressed Concrete Institute.

CSTR. (2005). *Post tensioned concrete floors design handbook. CSTR 43*. Camberley Surrey: Concrete Society.

Gales, J., Bisby, L., & Gillie, M. (2011a). Unbonded post tensioned concrete in fire: A review of data from furnace tests and real fires. *Fire Safety Journal, 46*(4), 151–163.

Gales, J., Bisby, L., & Gillie, M. (2011b). Unbonded post tensioned concrete slabs in fire – Part I – Experimental response of unbonded tendons under transient localized heating. *Journal of Structural Fire Engineering, 2*(3), 139–154.

Gales, J., Bisby, L., & Gillie, M. (2011c). Unbonded post tensioned concrete slabs in fire – Part II – Modelling tendon response and the consequences of localized heating. *Journal of Structural Fire Engineering, 2*(3), 155–172.

Gales, J., Bisby, L., & Stratford, T. (2012). New parameters to describe high temperature deformation of prestressing steel determined using digital image correlation. *Structural Engineering International, 22*(4), 476–486.

Gales, J., & Green, M. (2015). Optical characterization of high temperature deformation in novel structural materials. In *Proceedings of the 14th International Conference on Fire and Materials*, San Francisco, CA, pp. 626–640.

Hager, I. (2014). Colour change in heated concrete. *Fire Technology, 50*, 945–958.

Ghalleb, A. (2013). Calculating ultimate tendon stress in externally prestressed continous concrete beams using simplified formulas. *Engineering Structures, 46*, 417–430.

IBC. (2012). *International building code*. Country Farm Hills, Il., USA: International Code Council.

Ingham, J. (2009). Forensic engineering of fire-damaged structures. *Proceedings of the Institution of Civil Engineers: Civil Engineering, 162*(5), 12–17.

ISO 834. (1999). *Fire resistance test – Elements of building construction*. Geneva: International Organization for Standardization.

Khoury, G. A. (2006). Strain of heated concrete during two thermal cycles. Part 2: Strain during first cooling and subsequent thermal cycle. *Magazine of Concrete Research, 58*, 387–400.

Roberston, L., Dudorova, Z., Gales, J., Stratford, T., & Blackford, J. (19–20 April 2013). *Microstructural and mechanical characterization of post-tensioning tendons following elevated temperature exposure*. Applications of Structural Engineering Conference. Prague, pp. 474–479.

Chapter 5
Recommendations for Advancing the Fire Safe Design of Post-tensioned Concrete

> An ignorant person may make many mistakes without being
> aware that he has done so, and the slightest failure is probably
> fatal to everyone...
>
> —James Braidwood First fire master of Edinburgh and founder
> of the London Fire Brigade (Braidwood 1849)

5.1 Overview

As most jurisdictions move towards the adoption of performance-based structural fire design codes it is crucial that the behavior of all structures in fire is scientifically and rationally understood. Post-tensioned (PT) concrete structures are no exception. PT concrete is widely believed to benefit from its 'inherent fire endurance.' This belief, particularly for the unbonded configuration, is potentially problematic. It is based on results from standard fire tests performed on simply supported specimens more than five decades ago. The credibility of such tests is debatable. Not only are they unable to reflect the behavior of modern PT concrete construction materials, they cannot capture the true structural behavior of real buildings in real fires. The objective of this book has been to fill in some of the significant knowledge gaps regarding the structural and thermal response of PT concrete structures in fire, particularly the performance of the unbonded prestressing tendons in floor slabs. This knowledge is vital for the creation of defensible computational modelling of unbonded PT structures in fire.

A review of previous experiments and real case studies of PT concrete structures exposed to fire, as given in Chaps. 2 and 3, provides evidence that current code guidance for the fire-safe design of modern PT concrete slabs, particularly unbonded, are not adequate. While there are research gaps for both bonded and unbonded post tensioned (UPT) configurations, the dangers of the tendon rupture in unbonded configurations, due to localized heating, needs particular attention. The interaction between thermal relaxation and plastic deformation for unbonded prestressing

© The Author(s) 2016
J. Gales et al., *Structural Fire Performance of Contemporary
Post-tensioned Concrete Construction*, SpringerBriefs in Fire,
DOI 10.1007/978-1-4939-3280-1_5

tendons during a localized fire can result in failure considerably earlier than what can be predicted from available design guidance. This danger is increased in cases where concrete cover is reduced. Since unbonded prestressing tendons run continuously in UPT slabs, local damage to the tendon will affect the integrity of adjacent bays. In the event that no bonded steel reinforcement is provided, which is currently permitted by some design codes, a PT concrete slab could lose tensile reinforcement across an entire floor of the building. This occurrence has been seen in real building fires.

Structural actions within a real PT concrete structure, such as thermal bowing, material behavior, restraint, concrete stiffness loss, continuity, spalling, and slab splitting, could also play significant roles influencing the PT structure's response in fire. Three non-standard structural fire experiments were performed on continuous and restrained PT concrete slabs in an effort to better understand the response of PT concrete structures to localized high temperature exposure. These experiments validate the need for a computational stress relaxation model for tendons located inside a real concrete structure during heating. This was the first time a continuous PT slab under axial, vertical and rotational restraint has been studied at high temperature. UPT slabs were heated locally in the center span using a radiant panel array until the embedded prestressing steel reached a critical temperature. The structural response during all tests indicated a unique five phase deflection response different from that of a simply supported slab during a standard fire test. Deflection trends in the continuous slab tests were due to a complex combination of thermal expansion and plastic damage. These test results will enable future efforts to build and validate rational computational modelling capability which can account for the requisite complexities of this relatively simple structural system.

5.2   Design Recommendations

Currently, some design guidance for modern unbonded PT concrete structures may be inadequate. This guidance should prevent the sudden rupture of prestressing tendons in UPT buildings during real fires real, yet real fire case studies have shown it may not. The broader structural consequences of the loss of prestressing reinforcement remain largely unknown, and it is difficult to conclusively state whether other relevant structural interactions, such as membrane actions or moment redistribution, etc., may prevent a given PT slab from collapsing during or after a fire. These are questions that need to be asked and answered, and used to validate computational modelling in the future. Even without this additional research, it has been shown that updated minimums for concrete cover and bonded reinforcement for UPT concrete slabs are necessary in the existing international guidance.

The amount of concrete cover required by EN 1992-1-2 (CEN 2004) is based on the steel tendon having a critical temperature of 350 °C. While this may be defensible for simply supported slabs, is insufficient for preventing tendon rupture in cases of localized heating. Increasing the EN 1992-1-2 prescribed covers by even

5 mm could address this issue (see Gales et al. 2011 for discussion). The assumed critical temperature of 427 °C in North American design guidance (IBC 2012) is particularly difficult to justify and is in need of revision. In both EN 1992-1-2 (CEN 2004) and the IBC (2012) the required concrete covers for restrained slabs are even smaller than the unrestrained case. The reason given for this that restraint and continuity (redistribution of moments) aid in collapse prevention during. While this approach may be defensible on the basis of standard fire tests, it makes it very likely that tendon rupture will occur in restrained slabs well before the prescribed fire resistance time is met. At the very least the EN 1992-1-2 (CEN 2004) cover requirements should be universally adopted in North American codes, guidance and standards, assuming the premature tendon rupture is to be avoided during fire.

The assumption that slab failure will not occur during or after a fire in the event of tendon rupture, particularly for structures without minimum amounts of bonded reinforcement, is difficult to justify. Designers should be required to supply a minimum amount of bonded reinforcement in all UPT slabs to enable load shedding in the event of tendon rupture. The current situation, where some codes allow for no bonded reinforcement in certain regions of a slab, may lead to unexpected failures. While the recommended bonded reinforcement ratios of 0.2 % as prescribed by Van Heberghen and Van Damme (1983) should be sufficient for the time being, future research is needed to defensibly suggest the minimum amount needed. Additionally, a series of transverse ties could help prevent longitudinal cracking along the tendon in UPT slabs, if this is presumed to be problematic.

5.3 Research Needs

The tests presented in Chap. 4 represent the most simplistic experimental design that can account for as many relevant as-built construction features as possible. The slabs were designed to explore the behavior of PT concrete structures in fire, not to provide 'pass-fail' fire test criteria for prescriptive design. It is the authors' hope that the data from the tests presented in this book, along with other future tests, will eventually lead to modelling capabilities that enable designers to develop rational fire safety strategies for concrete buildings, PT or otherwise. Areas that require additional research attention to correctly account for the complexities in deformation include:

- **Mechanical re-stressing of unbonded tendons**—When exposed to heating, many inter-related structural mechanisms may influence the stress level of unbonded prestressing tendons. The roles of thermal bowing, loss of pre-compression, and restraint on the stress level of unbonded tendons need further consideration.
- **High temperature relaxation of prestressing tendons**—Deformation was shown to be heavily dependent on tendon stress relaxation due to thermal relaxation and creep elongation. Future work is required to define a tendon stress

relaxation model that can correctly account for variable heating rates. To address this need, a full tendon model has been constructed by the primary authors to appear in print in the near future.

- **Consequences of tendon rupture**—Research is needed in order to understand the effects of immediate load shedding to bonded (non-prestressed) steel reinforcement, and whether the structure can maintain the applied load after unbonded tendon rupture.
- **Concrete behavior in fire**—A detailed study into the thermal and mechanical behavior of concrete and concrete structures may aid modelling efforts in the future. A more thorough understanding of conventional concrete is required (Gales et al. 2015). Specifically, spalling, cracking, load induced thermal strains; stiffness loss, and material parameters in cooling, all need research to enable credible modelling of conventional concrete behavior.
- **Mild reinforcement material properties**—At ambient temperature, the mechanical properties of the mild steel were seen to be equivalent to other researchers. High temperature tests should be conducted to verify this performance. This will give the confidence needed in order to specify the appropriate material model for reinforcing steel in any numerical model.
- **Accurate model inputs**—Any high temperature model must adequately account for the initial conditions of the structural system. This includes the degree of pre-compression, forces exerted on the columns from post-tensioning, etc. This knowledge will aid in determining the degree of flexural/compressive prestress in the concrete at the start of heating, and allow for appropriate consideration of load-induced temperature effects.

5.4 Final Words

Unrealistic standard testing has left the true behavior of real UPT concrete structures in fire a relative mystery. While there are still many areas where additional research is needed, the tests covered in this book help shed light on what is clearly a complex behavior. The continuous, restrained, one-way spanning PT slabs tested exhibited five distinct phases of deflection under localized heating and cooling. These trends appear to be influenced by a complex interplay between stiffness degradation of the concrete, plastic deformation of the prestressing steel tendon, and differential thermal expansion and contraction. Attempts to credibly model these structural systems in fire must defensibly account for these, and other, complexities. The results from these tests in combination with future experimentation, which addresses the research needs given above, will allow for, and hopefully lead to, the development of rational fire safety strategies for optimized PT concrete buildings.

References

Braidwood, J. (1849). On fire proof buildings. *Proceedings of the Institution of Civil Engineers, 767*, 141–163.

CEN. (2004). *Eurocode 2: Design of concrete structures, Parts 1–2: General rules-structural fire design, EN 1992-1-2*. Brussels: European Committee for Standardization.

Gales, J., Bisby, L., & Gillie, M. (2011). Unbonded post tensioned concrete slabs in fire – Part II – Modelling tendon response and the consequences of localized heating. *Journal of Structural Fire Engineering, 2*(3), 155–172.

Gales, J., Parker, T., Cree, D., & Green, M. (2015). Fire performance of sustainable recycled concrete aggregates: Mechanical properties at elevated temperatures and current research needs. *Fire Technology*, 29 pp. In press.

IBC. (2012). *International building code*. Country Farm Hills, Il, USA: International Code Council.

Van Herberghen, P., & Van Damme, M. (1983). *Fire resistance of post-tensioned continuous flat floor slabs with unbonded tendons*. FIP Notes, UT, pp. 3–11.

Appendix: Reference Data

The tables presented in this section chronologically list the 46 standard fire tests of unbonded PT concrete assemblies available in publically accessible peer-reviewed literature. The tables provide details on the following testing parameter details: year, assembly description, restraint condition, precompression, bonded steel, moisture content, aggregate type, strength of concrete at testing, and load ratio. Also highlighted are significant test observations including maximum tendon temperature, span to depth ratio, observed cracking and spalling, tendon rupture occurrence, and test termination criterion. A more exhaustive description of tests 1–14, 20–28 and 38–41 can be found in *Unbonded Post Tensioned Concrete in Fire: A Review of Data from Furnace Tests and Real Fires* within the *Fire Safety Journal*. Other tests are discussed within Chap. 3 of this book.

© The Author(s) 2016   85
J. Gales et al., *Structural Fire Performance of Contemporary
Post-tensioned Concrete Construction*, SpringerBriefs in Fire,
DOI 10.1007/978-1-4939-3280-1

Table A.1 Unbonded post tensioned concrete furnace tests

Part Ia

#	Year[a]	General description	Concrete precompression due to prestress (longitudinal, MPa/ transverse, MPa)[b]	Restraint conditions (lateral/rotational)[c]	Specimen pre-condition[d]	Bonded steel reinf.[e] (%)	Moisture content at testing	Aggregate type
1	1958	Two-way beam-slab assembly	0.94/1.13	Y/P	–	0/0/0/0	62 % RH	Siliceous
2	1959	Two-way slab panel	1.64/1.61	Y/P	–	0/0/0/0	76 % RH	Siliceous
3	1964	T-beam	6.85/0	N/N	Y	Mesh in cover	75 % RH	Carbonate
4			6.08/0				72 % RH	Carbonate
5	1965	Inverted T-beam	3.29/0	Y/P	–	0/0/0/0	–	Carbonate
6	1967	Two-way slab panel	2.39/2.07	Y/P	Y	0/0/0/0	44 % RH	Expanded shale
7	1983	Continuous one-way slab strip	0.78/0	N/Y	–	0/0/0/0	–	Gravel
8			0.78/0			0/0/0/0		Gravel
9			0.78/0			0/0/0.12/0.12		Gravel
10			0.78/1.65			0.26/0.26/0.1/0.1[f]		Gravel
11			0.78/0			0.15/0.15/0.08/0.08		Limestone
12			0.78/0			0.2/0.2/0.1/0.1		Limestone
13			0.78/0			0.2/0.2/0.1/0.1		Gravel
14			0.78/0			0.2/0.2/0.2/0.2		Gravel

Part Ib

#	Year[a]	fc' at testing (MPa)	Load ratio[g]	Max tendon temp. (°C)	Span/depth ratio	Longitudinal crack?	Transverse crack?	Spalling?	Tendon rupture?	End point
1	1958	41	1.0×LL	506	21	Y	Y	Y	N	Transmission
2	1959	30	1.0×LL	563	28	–	–	–	–	None
3	1964	39	0.516	541	–	Y	N	N	N	Collapse
4	1964	41	0.520	427	–	N	Y	Y	N	Collapse
5	1965	–	1.0×LL	377	–	N	N	N	N	None
6	1967	29[h]	1.0×LL	704	23	N	Y	N	N	None
7	1983	–	Varies	–	33	Y	Y	Y	Y	Collapse or imminent collapse
8										
9										
10										
11										
12										
13										
14										

(continued)

Table A.1 (continued)

Part IIa

#	Year[a]	General description	Concrete precompression due to prestress (longitudinal, MPa/transverse, Mpa)[b]	Restraint conditions (lateral/rotational)[c]	Specimen pre-condition[d]	Bonded steel reinf.[e] (%)	Moisture content at testing	Aggregate type
15	2004	Continuous one-way slab strip	–	N/P	Water reducer	Mesh in cover	–	Gravel
16			0.54					
17			0.58					
18			0.59					
19	2005	Continuous one-way slab strip	1 to 0.2/0	N/P	–	–	–	–
20	2006	One-way slab strip	0.77/0	N/N	–	0/0.45/0.21	4.0 % by wt	Carbonate
21			0.98/0			0/0.31/0.21	3.5 % by wt	
22			2.09/0			0/0.31/0.21	4.0 % by wt	
23			0.88/0			0/0.46/0.18	2.4 % by wt	
24			1.09/0			0/0.31/0.18	3.3 % by wt	
25			1.19/0			0/0.21/0.18	2.4 % by wt	
26			0.7/0			0/0.25/0.17	3.3 % by wt	
27			0.96/0			0/0.29/0.17	3.5 % by wt	
28			1.44/0			0/0.20/0.17	1.8 % by wt	

Part IIb

#	Year[a]	fc′ at testing (MPa)	Load ratio[g]	Max tendon temp. (°C)	Span/depth ratio	Longitudinal crack?	Transverse crack?	Spalling?	Tendon rupture?	End point
15	2004	48.6	1.0×LL	200	33–35	Y	Y	–	N	None
16		58		125	33–35	Y	Y	Y	Y	
17		51.6		325	33–35	Y	Y	–	N	
18		55.8		350	33–35	Y	Y	–	N	
19	2005	–	–	–	23	–	Y	N	N	None
20	2006	57	0.41	–	41	N	Y	N	N	Slope limit
21		48	0.54		41	Y	Y	Y		
22		57	0.70		41	N	Y	Y		
23		52	0.42		37	N	N	N		
24		52	0.56		37	Y	N	N		
25		52	0.68		37	N	N	N		
26		52	0.42		35	N	Y	Y		
27		47	0.56		35	N	Y	Y		
28		23	0.72		35	N	Y	Y		

(continued)

Table A.1 (continued)

Part IIIa

#	Year[a]	General description	Concrete precompression due to prestress (longitudinal, MPa/ transverse, MPa)[b]	Restraint conditions (lateral/rotational)[c]	Specimen pre-condition[d]	Bonded steel reinf.[e] (%)	Moisture content at testing	Aggregate type
29	2007	Continuous one-way slab strip	0.92/0	N/P	–	1.2/0.18/0.61/0.10	3.98	Carbonate
30			1.15/0			1.2/0.18/0.39/0.10	3.98	
31			1.42/0			1.2/0.18/0.23/0.10	3.5	
32			0.75/0			1.1/0.15/0.41/0.09	2.77	
33			1.19/0			1.1/0.15/0.41/0.09	2.36	
34			1.24/0			1.1/0.15/0.21/0.09	3.5	
35			0.76/0			1.0/0.15/0.50/0.08	2.36	
36			0.86/0			1.0/0.15/0.29/0.08	3.29	
37			1.27/0			1.0/0.15/0.22/0.08	3.29	
38	2008	One-way slab strip	1.97/0	N/N	N	0/0/0/0	2.5	Limestone
39			1.97/0	Y/P			2.2	
40			1.97/0	N/N			2.3	Thames gravel
41			1.97/0	Y/P			1.7	
42	2011	Bi-directional panel	2.04/2.87	N/N	–	2.4	4.7	–
43	2011	One-way slab strip	1.2/0	N/N	N	0/0/0/0	–	–
44			1.0/0	N/N			5.7	–
45	2012	Continuous one-way strip	2.37/0	Y/Y	N	0.22/0/0.22/0	4.0	Mixed gravel
46			2.39/0				3.8	

Part IIIb

#	Year[a]	fc' at test (MPa)	Load ratio[g]	Max tendon temp. (°C)	Span/depth ratio	Longitudinal crack	Transverse crack?	Spalling?	Tendon rupture?	End point
29	2007	56.9	0.36	–	36	–	–	N	N	None
30		56.9	0.5			–	–	N	N	
31		48.2	0.59			N	Y	Y	Y	Critical temp
32		40.1	0.58			–	–	N	N	Critical temp
33		52.1	0.33			–	–	N	N	Critical temp
34		48.2	0.36			N	Y	Y	Y	Critical temp
35		52.1	0.42			–	–	N	N	
36		52	0.48			–	–	N	N	
37		52	0.32			–	–	Y	N	
38	2008	48	0.65	492	23	Y	N	N	Y	Collapse
39		41		350			Y	Y	N	Critical temp
40		40					N	N	N	Critical temp
41		40					N	N	N	Critical temp
42	2011	38.7	–	587–718	5	Y	Y	Y	N	RABT
43	2011	63	–	~150	15	?	?	N	N	None
44		54			12					
45	2012	41	0.42	361	43	N	Y	N	N	Critical temp
46		43		432		Y	Y	N	N	Critical temp

[a]Refers to the year the tests were conducted (estimated on the basis of publication date in some cases)

[b]The precompression is the total tensioning force divided by the cross-sectional area of the slab normal to that force (Taranath 2010)

[c]Y yes, N no, P partial (assumed by current authors)

[d]"–" means that this information was not disclosed

[e]Top longitudinal/top transverse/bottom longitudinal/bottom transverse

[f]This slab was also prestressed transversely

[g]In some cases the load ratio is not given, in which case the loading approach is given instead

[h]28-Day compressive strength (strength at time of testing not known)

Printed in the United States
By Bookmasters